GIANT REPTILES

By the same authors

Venomous Reptiles

GIANT REPTILES

Sherman A. Minton, Jr., and
Madge Rutherford Minton

CHARLES SCRIBNER'S SONS · NEW YORK

Library of Congress Cataloging in Publication Data

Minton, Sherman A
Giant reptiles.

Bibliography: p.
1. Reptiles. 2. Serpents (in religion, folk-lore, etc.) I. Minton, Madge Rutherford, joint author. II. Title.
QL641.M5 598.1 72-9770
ISBN 0-684-13267-2 (cloth)
ISBN 0-684-14558-8 (paper)

This book published simultaneously in the United States of America and in Canada—

3 5 7 9 11 13 15 17 19 V/C 20 18 16 14 12 10 8 6 4 2
1 3 5 7 9 11 13 15 17 19 V/P 20 18 16 14 12 10 8 6 4 2

Printed in the United States of America

For

Bette and Bern
Mickey and Chuck
Chugtai and Gil
Dorothy and Phil
Max, John, Harry, Frank
Lauretta and Jerry

Contents

Illustrations

Preface

This book is about the giant reptiles that have lived into the time of man, influenced his thinking, and contributed to his economy and culture. They have representatives in every order of reptiles alive in the world today, except for one that has been reduced to a single species. In comparison with human history, the groups of reptiles are incredibly old. The snakes have some 75 million years of evolution behind them, and the other groups are more than twice that old, going back to the early Mesozoic, when the reptiles began to utilize to the fullest extent their capacity for diversity and great size. This trend climaxed in the dinosaurs, and when they disappeared, reptiles declined spectacularly in size and variety. Nevertheless, among the survivors are some of the largest terrestrial and amphibious animals that still inhabit the globe. While man gained his humanity in a world inhabited by woolly mammoths, giant sloths, mastodons, saber-tooth cats, cave bears, and a number of other huge mammals now extinct, the big reptiles of that world were, with very few exceptions, the same species here today. There is no zoological evidence that human eyes have seen reptiles much greater in size or different in form than those alive at the beginning of the twentieth century. The Biblical Leviathan, the dragons of China and Japan, the giant lizards of Polynesia, the great snakes of European, Asian, and Australian mythology, and the cosmic turtles of India and pre-Columbian America can all be traced directly to animals that still exist.

Whether these reptiles that man once regarded as supernaturally powerful will exist much longer is now very largely an option for man to exercise in the next few decades. Today Leviathan has been all but drawn out of the world's waters; his skin is full of barbed irons and his head pierced with fish spears. The decline in numbers of the world's giant reptiles, especially the turtles and crocodilians, has been precipitous in the last century, catastrophic in the last few decades. Some species such as the Cuban crocodile and Chinese alligator may already be past salvage; others can be saved only by determined efforts that in many cases must be on an international scale. Captivity or semidomestication may be the ultimate lot of some species that are economically valuable and will adapt to such a regimen. Most will retreat to the few remaining wild places, there to remain until deterioration of the environment and pressure from growing human populations deracinate them. A very few may adapt to life along the fringes of civilization and persist in spite of man rather than because of his protection. The reasons for the decimation of the giant reptiles are many; we touch on some of them in the chapters to follow. Their plight is literally a *Götterdämmerung*, for gods they have been to many peoples of many lands.

As we did in our previous book, *Venomous Reptiles*, we have tried here to summarize factual information and relate it to the rich store of reptile myth and folklore. Reptile legends are as widespread as they are imaginative. It is surprising to hear the same story of snakes sucking milk from the udders of cows in Floyd County, Indiana, and in the Tatta District of Sind, and to see a woodcarving from New Guinea illustrating a crocodilian myth similar to one Christopher Columbus heard from Indians of the Caribbean.

Our definition of a giant reptile is personal and arbitrary. Since all but four of the crocodilians reach lengths of 6 feet or more, we have considered the whole group giants, although we have not attempted to give space to all the species. Among the turtles our definition includes any species with a shell length of 2 feet or more or a weight of over 50 pounds. A giant snake is one reaching a length of 10 feet or more, and a giant lizard one reaching a length of 6 feet or more. Again, we have not given comparable space to all species. While admitting the advantages of the metric system of weights and measures and advocating its worldwide adoption, we feel that most Americans are still more at home with the English system, and accordingly we have used it in this book. If our original information source used metric units, we have retained them but provided approximate English equivalents.

In the text we have referred to the reptiles by English or other vernacular names except in a few cases where no suitable name exists. Scientific names are listed in the Appendix.

Both of us have personal stories to tell and, of necessity, we have freely used the first person singular pronoun. In chapters largely devoted to mythology and religion, "I" refers to Madge, while in chapters concerning biology, ecology, and distribution, Sherman speaks.

The giant reptiles fascinate us, as they do most people. Although our encounters with them in the wild have been brief, they have usually been memorable. Etched forever in our minds are the sight of our first wild alligators along the Tamiami Trail in the Everglades, the red glow of crocodile eyes in pools of the Hab River in Baluchistan, sea turtles heaving their gray-green bulks up the beaches of the Arabian Sea to lay their eggs, and the deep, straight track of a big python in tall grass near the Indus. We hope that such experiences will not end with our generation. Sometimes man needs to feel the presence of an older, alien living world. Even the slight but real menace of these animals may not be a wholly bad thing. Man grows increasingly uneasy in a world where danger has only a human face.

GIANT REPTILES

1. Dinosaurs and Other Vanished Giants

The sun burned in a cloudless blue Texas sky. Its heat bounded off the reddish granite boulders of the Rosillos Mountains and onto a little plateau. Except for some scattered clumps of cactus and creosote bush, there was not a green plant in sight. But it had not always been so. Stumps and chunks of tree trunks, some of them 30 feet long, lay all around. Their wood had long ago been replaced molecule by molecule with blackish jasper occasionally streaked with scarlet and orange. In this sun-scorched spot Madge and I were realizing a childhood dream. We were digging up a dinosaur. With plaster of Paris and burlap, Madge was putting a cast on a yard-long chalky bone weighing about 20 pounds. Our eldest daughter, Brooks, and I were picking and brushing the hard soil away from other bones. Two younger daughters toddled about, busily but aimlessly, occasionally picking up pieces of jasper and licking them to bring out their colors.

Dinosaur bones were no novelty in the Big Bend. "Damn big ox bones," said one of our neighbors, who had several of them lining the path in front of his house. The bones we were digging had been shown us by two young ranch hands who, because schooling had come slowly to that remote part of Texas, were in the fifth grade with Brooks. Like the rest of the local folk, they regarded the bones as one more mildly interesting oddity in a land that was long on natural wonders but short on cash and rainfall.

Judging from what appeared to be a short, wide humerus and several spool-shaped carpals and metacarpals, our find was the flipper and shoulder region of a plesiosaur or some similar large aquatic reptile. Its bones were probably several million years older than the fossilized trees, dating back to a time when shallow seas covered much of North America. Fossils were not our major reason for being in the Big Bend, so our dinosaur dig lasted only a few days. When we finally returned to Indiana with close to a hundred pounds of fossil bone, we made another interesting discovery about dinosaurs. Their bones are a drug on the scientific market. Museums that eagerly snapped up the mice, lizards, snakes, and scorpions we had collected said, "Thanks, but no thanks," to the dinosaur remains—"unless perchance you have a skull."

The reptile whose bones we found had been quite literally an inhabitant of a different world. An interplanetary voyager circling the earth during the Triassic would have seen a pattern of land and sea totally unlike the one Yuri Gagarin and John Glenn glimpsed from their spinning satellites 200 million years later. There were then two great land masses or supercontinents, the southern Gondwanaland (named from the Gondwana province of India) and the northern Laurasia (from the Laurentian region of Canada plus Asia), separated by the Tethys Sea whose remnants today are the Mediterranean and the Persian Gulf. The Atlantic Ocean did not exist, and the Pacific was much larger than at present. The Triassic was a critical period in reptile evolution. Most of the earlier Permian reptiles had disappeared, but some had set foot on the path that was soon to lead to the first mammals. Some fifteen other major groups of reptiles, including the dinosaurs, rhynchocephalians, turtles, crocodilians, and lizards, made their appearance. Fossil evidence indicates that Gondwanaland was an active center for reptile evolution, while more primitive types tended to persist in Laurasia.[1] Today a contrary situation exists—the surviving primitive reptiles and mammals are largely confined to the southern continents and islands.

At the end of the Triassic (about 168 million years ago) Gondwanaland began to fragment, beginning when a land mass destined to become Antarctica and Australia broke away and began drifting eastward, while the Indian peninsula sailed off to the north. Somewhat later South America began to pull away from Africa, and an arm of the sea began to separate North America from Europe. Late in the Cretaceous, less than 100 million years ago, Madagascar and some fair-sized islands whose remnants are the present-day Seychelles split off from Africa, Australia cracked free of Antarctica and shifted northeastward, and the Indian peninsula ended a voyage of some 5600 miles by plowing into the Asian mainland. Meanwhile, Europe and North America were pulling apart with the formation of the North Atlantic, while the widening gap between Africa and South America became the South Atlantic.[2] The spread of land reptiles such as the giant dinosaurs from southern to northern continents during the Mesozoic period was faster than can be accounted for by the movements of the continents themselves and provides strong evidence that land bridges or island chains linked parts of Laurasia and Gondwanaland at least intermittently.

For the past 40 million years, the continents have had more or less their present positions, although there has been a good deal of rearrangement of shorelines corresponding with rises and falls in sea level. Land bridges connecting Asia and North America in the region of the present-day Aleutian Island chain have been very important in allowing Asian animals access to America. For much of its history, South America has been separated from North America, sometimes by as much as 300 miles. Central America has alternately been an isthmus and a chain of islands; the present fusion of the continents has probably existed no more than 5 million years.

In the strict sense, dinosaurs are members of two orders of terrestrial or amphibious reptiles that existed from the early Triassic to the end of the Cretaceous, a period of about 100 million years, during which time they became numerous and highly diversified and were the dominant land vertebrates. In popular usage, the term has been expanded to include certain other groups of large reptiles, such as the marine ichthyosaurs and plesiosaurs which were more or less contemporary with the dinosaurs, although they were anatomically quite different and arose from different stock. Nearly all the dinosaurs were giant reptiles by contemporary standards. One of the smallest known species, *Compsognathus*, was the size of a small chicken, which is considerably larger than the vast majority of modern lizards. The most primitive dinosaurs were light-bodied creatures that ran on their hind legs. They apparently were active predators and must have been somewhat like roadrunners or secretary birds in their habits.

In the late Triassic and the Jurassic, probably along with the development of vast swamplands and an inexhaustible supply of succulent vegetation, a group of dinosaurs evolved that walked on four legs, had long necks and correspondingly long tails and comparatively small heads with short spoonlike or peglike teeth. In this group, technically known as the sauropods, are the largest land animals the world has ever seen. Endlessly and usually anachronistically depicted in books, cartoons, movies, and advertisements for petroleum products, they are as familiar to much of the world's population as many animals that still exist. *Brontosaurus* (or *Apatosaurus* if you are a purist in taxonomic nomenclature) is the best known of these giants. It was about 65 feet long when full grown, stood 15 feet high at the hips, and weighed in the neighborhood of 40 tons. *Diplodocus* was even longer—close to 90 feet—but more slenderly built. It must have looked something like an immense python that had swallowed a Volkswagen and sprouted four stumpy legs.

Brachiosaurus may have been the giant among giants. The largest reasonably complete specimen is about 70 feet, but fragmentary remains suggest that some individuals may have been still larger. This sauropod was heavily built and was unique in having high forequarters and comparatively low hips. Like a colossal giraffe, it must have browsed some 40 feet in the treetops. A remarkable series of fossilized footprints found in a Texas streambed demonstrated that giant sauropods walked about on land reasonably well and probably traveled in groups of a dozen or two. This same trackway indicated that the sauropods were being followed by a giant bipedal carnivorous dinosaur, doubtless waiting for a young or weak animal to drop behind the rest. Other tracks indicate that sauropods did a good deal of wading in water deep enough to support much of their weight, as indicated by the absence of tail tracks.[3] It is doubtful if they were good swimmers. A recent paper speculates that they may have been primarily terrestrial. The skeleton of the legs, chest, and tail is that of a land animal rather than an aquatic one, and the long neck could be associated with high browsing as readily as with feeding on aquatic plants. Sauropods may have lived in flood-plain forests where they consumed huge amounts of underbrush and helped keep the woodland relatively open.[4]

While tooth and jaw structure indicate that the sauropods were herbivores, the first direct evidence was obtained when William L. Stokes examined the fossilized stomach contents of one of these dinosaurs. The bulk of the material was twigs and branches chopped into pieces no more than an inch long and mixed with amorphous material that originally may have been mud. Most surprising was the presence of numerous pieces of bone, most of them tiny, and the tooth of an *Allosaurus*, a large carnivorous dinosaur.[5] This suggests that the sauropods may have been given to chewing on skeletons or carcases, perhaps to obtain supplemental calcium and other minerals. Incidentally, coproliths—fossilized manure—of dinosaurs are not particularly rare and can be obtained from lapidary or curio dealers for conversation pieces or may be cut and polished to make cuff links and tie clips.

The great carnivorous dinosaurs are nearly as well known as the sauropods. They walked on their hind limbs, and in some the forelimbs were so small as to be almost vestigial. Their skulls were large and strongly but lightly constructed, the bones providing attachment for mighty jaw and neck muscles. Their teeth were serrate daggers admirably adapted for cutting and tearing. Like snakes they had a large gape and probably bolted their food in huge chunks. *Tyrannosaurus*, the best-known member of this group, was one of the last dinosaur species to appear before the great extinction at the end of the Cretaceous. It was nearly 50 feet long and may have been able to rear up to a height of 20 feet, although its normal posture was probably closer to the horizontal. A few footprints and isolated bones suggest that there may have been bigger carnivores, but, like the giant sauropods, these must have been approaching the limits of supporting and locomotor power inherent in bones, ligaments, and muscles.

The great marine reptiles, curiously enough, never reached the sizes attained today by their successors, the whales. Perhaps they lacked either the rich food supply or the means to harvest it which have enabled the whales to become the largest animals to inhabit the earth. The fishlike ichthyosaurs were generally less than 10 feet long, although some may have approached 30 feet. The plesiosaurs produced some large species, such as *Elasmosaurus* which reached a length of 43 feet, more than half of it neck. *Kronosaurus*, a whalelike offshoot of the plesiosaur group, reached about the same length but was a more massive animal.[6] The mosasaurs were snakelike marine reptiles with short paddlelike limbs. Their largest species also approached the 40-foot mark. The mosasaurs have a fairly close affinity to the modern monitor lizards. At a recent Indiana State Fair a large monitor lizard exhibited on the midway was billed as "The Mosasaurus of the Jungle."

Learning much about the life histories of animals dead 70 million years sounds fairly hopeless, but intelligent inferences plus a few lucky finds can go a long way. Nevertheless, many questions are unanswered and probably always will be. Very little is known of secondary sexual characters in Mesozoic reptiles, although there is a good chance that some, such as spurs or horns, have been fossilized. The location and form of sex organs themselves can only be guessed by comparisons with living reptiles. Copulation must have presented difficulties in some of the more heavily armored and massive dinosaurs, but they managed somehow.

That some dinosaurs laid eggs there is no doubt. An American Museum of Natural History expedition led by Roy Chapman Andrews discovered many eggs and some virtually undisturbed nests of *Protoceratops*, a small, stocky dinosaur with a neck frill. Its eggs were oval and about 9 inches in greatest diameter. They were deposited in concentric circles, evidently in a shallow hole scooped in sand. The finding near one nest of a predacious dinosaur with its skull crushed suggests that *Protoceratops* may have guarded its eggs. The only other dinosaur eggs that have been found in numbers and can be associated with a particular species are those of *Hypselosaurus*, a sauropod about two-thirds the size of *Brontosaurus*. These eggs are more nearly spherical and almost twice the size of an ostrich egg, or about 11 inches in greatest diameter. Although some have been found in groups of five, there is little evidence of how and where they were deposited.[7] It would seem difficult for the sauropod dinosaurs with their elephantine feet to dig a hole for their eggs as turtles do. Perhaps they were able to rake masses of decaying vegetation over the eggs, constructing a mound nest such as is built by the king cobra and the alligator.

No conclusive evidence that any dinosaurs bore live young has been obtained. In modern snakes and lizards the female's ability to retain the eggs within her body until the embryo is fully developed is well known, and some species do not seem to be firmly committed to either mode of reproduction. The saw-scaled vipers (which may comprise two or more species) in Israel and parts of Africa lay eggs but in India and Pakistan are live-bearing. The vivaparous lizard is live-bearing over most of its range, but a population in the Pyrenees lays eggs. The marine ichthyosaurs were unquestionably live-bearing; females containing fossilized embryos are known, and one specimen evidentally died and was fossilized in the act of giving birth.[8] The mosasaurs and some plesiosaurs were most probably live-bearing, since their body form was such that it would have been very difficult for them to leave the water to deposit eggs. Other plesiosaurs probably could have hauled out on the beach like sea turtles and buried their eggs in the sand.

Young dinosaurs are remarkably rare in the fossil record, although immense numbers of hatchlings must have been produced each season. No doubt most of them ended up in the stomachs of other dinosaurs; moreover, their bones, being small and probably lightly calcified, were not likely to be preserved. The scarcity of older juveniles may be partly explained by assuming that growth from hatching to adult size was rapid, followed by a slowdown and a prolonged period of relatively secure maturity and old age. The immature individuals would therefore have been a relatively small segment of the population. Something like this seems to happen today in sea turtles, where hatchlings and large adults are well known but half-grown specimens are less frequently seen.

The sudden extinction of the dinosaurs and several other groups of giant reptiles at the end of the Cretaceous, some 70 million years ago, is one of the most baffling events in natural history. They were so prolific and so highly diversified that it is hard to imagine either a catastrophe or a gradual change that would have destroyed them without eliminating all reptiles. However, some significant changes were occurring during the latter part of the Cretaceous. The primitive ferns, cycads, and conifers were being replaced by modern broad-leaved trees, grasses, and flowering plants. A period of mountain building, the Larimide Revolution, had begun to cause a gradual shift from tropical and subtropical maritime climate to climates with more strongly marked seasons. But these changes came very slowly, and there is good evidence that the big reptiles were adjusting to them. Although the ichthyosaurs and some families of dinosaurs disappeared about the mid-Cretaceous, other families, such as the herbivorous, terrestrial, horned dinosaurs typified by *Triceritops*, were increasing in numbers and variety of species; duckbill dinosaurs flourished; and the great carnivores reached their apogee in *Tyrannosaurus*. Mosasaurs and plesiosaurs were numerous in the seas, and pterosaurs were being at least as successful in the air as the primitive birds. And yet, within a couple of million years—a mere weekend as geological time goes—all the dinosaurs and most of the other big reptiles disappeared completely. Large size, which had been an asset for 100 million years, suddenly and mysteriously became a fatal liability. Size, however, was not the sole factor, for many small types of reptiles died off at about the same time. Of the major reptile groups that survived all still exist today. The lizards had comparatively little difficulty, although they lost the mosasaurs, their only family of real giants. The snakes, a group in its infancy, slipped through without trouble. The crocodiles suffered heavy attrition. Of the five crocodilian families known in the late Cretaceous, only one remained, and it lost many of its genera. Of the primitive rhynchocephalians, only the tuatara survives. Perhaps as early as 120 million years ago it retreated to New Zealand, where, in a favorable climate and without serious enemies or competitors, it waited out the millennia. The white man and the pigs, rats, cats, and other mammals he brought with him put an end to this peaceful state of affairs and very nearly to the tuatara, but it still exists under strict protection on a few offshore islets. The turtles suffered a considerable reduction and lost some spectacular forms, such as the huge marine *Archelon*, but retained their viability as a group.

What factor or factors caused the great extinction? Those who saw the Walt Disney film *Fantasia* may remember the sequence showing the dinosaurs plodding to their doom under a merciless sun, and climatic change of some sort has entered into the most generally held hypotheses accounting for the wholesale extinction of the Cretaceous reptiles. But all evidence indicates that the climatic changes at the time were neither sudden nor marked. They might have produced redistribution and local faunal changes but not a wholesale extinction. Competition between the reptiles and mammals seems out of the question—there were simply no large mammals around to compete. It has been suggested that small mammals killed off the dinosaurs by eating their eggs. There is no direct evidence of this (such would be hard to obtain), but this sort of predation, except on islands or in other restricted situations, leads to a balance in which both predator and prey continue to survive. Furthermore, this does not explain the demise of the marine reptiles.

Devastating, widespread disease has been suggested as an explanation, but this is contrary to present knowledge of epidemiology. Bacteria, viruses, and other pathogenic organisms generally have a restricted range of hosts. It is hard to imagine a newly evolved microbe that could produce fatal illness in so many more or less unrelated species. Parasitism also tends to strike a balance in which infection may be common, but acutely fatal disease is comparatively rare. Another suggestion, appropriate in a pollution-conscious age, is that fluctuation in the air's oxygen content or increase in some toxic trace element was the lethal factor. This has some plausibility but lacks supporting evidence. Extraterrestrial events such as solar flares that might have doused the world with radiation and sterilized or impaired the fertility of the dinosaurs and other reptiles have been evoked. Again there is little evidence for such events, and furthermore one would expect the germ cells of mammals to be more susceptible to such irradiation than those of reptiles.[9] It would seem that the great Cretaceous extinctions must have been the result of multiple factors, individually unimportant but widespread and persistent. Also, during the long and climatically bland Jurassic and early Cretaceous, many groups of reptiles, while retaining the capacity for great morphological change, lost their physiological adaptability. They became irrevocably committed to certain mechanisms, perhaps involving temperature regulation or reproductive biology, for example, and could not compensate for even minor environmental changes.

There have been other great dyings. At the end of the Permian period, some 230 million years ago and just before the beginning of the Mesozoic, nearly half the known families of all animals disappeared. Recovery of the world fauna to normal variety took 15 million to 20 million years. Of the primitive Permian reptiles, some 80 percent of the known families became extinct. Even less is known of the possible causes of this mass extinction. The Permian was a time of mountain building, during which the Ural, Appalachian, and Ouachita ranges were formed. It was also a time of climatic instability, with extensive glaciation in both hemispheres and the development of deserts.[10] We now seem to be living at the end of an era that has seen the disappearance of many large mammals. Living elephants and rhinoceroses are the last remnants of groups that were much larger and more diverse a few million years ago. The huge armadillo-like glyptodonts, the sabertooth cats, and the giant sloths are among the groups that have gone, although the last-mentioned survived in South America well into the time of man and may have been kept in semidomestication by the earliest Indians. Man has undoubtedly helped harry these big mammals into oblivion, but most were on their way out before his influence became decisive.

2. Introducing the Crocodilians

The nearest thing to a dinosaur that anyone can see alive in the wild today is an alligator or crocodile, but one had better not postpone the opportunity too long. These big reptiles are rapidly going the way of the dinosaur. Some may remain for decades in zoos and national parks or on ever-decreasing game preserves, but their role in shaping and balancing the ecology of wide areas is nearly at an end.

About the middle of the Triassic, some 180 million years ago, the first true crocodilians appeared in South America as a branch from the stock that eventually gave rise to the dinosaurs, pterosaurs, and birds. During the next 70 million years or so, crocodilians became quite diversified. The Jurassic thalattosuchians tried seagoing life, with some species developing very elongated bodies and flipperlike limbs and sacrificing much of their armor plating. The sebosuchians that lived in South America in the Cretaceous had high, narrow skulls and teeth like steak knives. They may have been forest animals rather than semiaquatic.[1] Most of the other prehistoric crocodilians appear to have closely resembled their living descendants. Some, like *Phobosuchus*, were enormous brutes 50 feet or so long and doubtless preyed upon the smaller dinosaurs.[2] *Stomatosuchus*, which lived in Africa in the late Cretaceous, was another gigantic species, but it had a flat, ducklike snout, many tiny teeth in the upper jaw, and none at all in the lower. Millions of years later, in the Miocene, another smaller but similar group of duckbilled crocodilians, the nettosuchians, made a brief appearance.[3] Probably these reptiles fed on small fish and other aquatic creatures by the technique sometimes known as "slurping"—that is, suddenly opening the wide mouth and creating a negative pressure that sucks the prey in.[4] The mata-mata turtle, the Surinam toad, and many fish feed in this manner, but it has evidently not proved particularly successful in the crocodilian line.

With very few exceptions, the meeting place of land and water has always been home for the crocodilians. Drenched with sunlight and richly supplied with plant and animal life drawn from both water and land, these areas are cornucopias for the large amphibious predator. The footing is too uncertain for the big meat-eaters of the land, while the big fish are in danger of stranding or becoming trapped in pools that became fatally depleted of oxygen. Independent of the water when he has to be but able to cruise it for prey or escape into it from enemies, the crocodilian makes the best of two worlds.

The two dozen or so surviving species of crocodilians are divided into three or four major groups that some herpetologists consider subfamilies and others families. One consists of a single, unique species, the gharial or gavial. Its most distinctive feature is a long, extremely narrow, and rounded snout. The jaws contain four rows of twenty-five to thirty interlocking teeth—more than in any other living crocodilian—and the skull bones are different in shape and arrangement from those of other crocodilians. The head resembles the long-necked Indian clay water jar, or *ghara*, which accounts for one of the animal's common local names, although this was long ago corrupted to gavial in European literature. The gharial is one of the largest of reptiles. Lorenze Hagenbeck of the famous German family of wild-animal hunters and dealers cited the account of a friend who killed a 30-foot specimen whose bloated carcass suggested a stranded whale.[5] Most recent authorities discount this record and give the maximum length as 20 to 23 feet. Nonetheless, fossils of a gharial 60 feet long have been found in India.

The gharial lives in the great rivers of India, Pakistan, and upper Burma. The people of Harappa and other ancient cities of the Indus Valley knew it well and depicted it on their carved seals, but it is not abundant in West Pakistan today. While Madge and I were in that country, we came to know the German herpetologist and collector, Mustapha G. Koniecznyi. He was a delightful conversationalist and served excellent beer, but in the field we were bitter rivals. One of his greatest coups was getting the skull of a gharial 6 meters (about 19.5 feet) long that had been killed along the Sutlej River in Bahawalpur District. To counter this, we tried to get a live gharial that had been caught by a fisherman living near Tatta in Sind and even made arrangements with an airline for shipping the beast alive to the United States, but in the meantime the fisherman sold his prize to a hide buyer.

Our friend Naim Beg Chugtai, a largely self-trained Pakistani naturalist and collector, told us that gharials could sometimes be seen during the winter months sunning on sand banks along the East Nara, a former channel of the Indus that now ends blindly in desolate salt flats of the Rann of Kutch. Late one afternoon, near the ancient Indus city of Sehwan, I saw the unmistakable eye-knobs of a crocodile gliding through the water near the mouth of a large canal. The body, mostly hidden in the muddy water, seemed too long and graceful for the Indian crocodile or mugger, and I like to think it was a gharial. If it was, it was probably the only one I shall ever see in its native home.

Historical accounts mention gharials as once common in the Ravi River near the city of Lahore, and another of our friends and field associates, Jerry Anderson, a Scotsman with many years' experience collecting and observing wild animals in Pakistan, obtained a few specimens from this stream. This section of Pakistan is at about the same latitude as Savannah, Georgia, and has three months of rather chilly weather. Evidently the gharial is fairly tolerant of cool temperatures, for it was able to thrive along the Ravi until hunters decimated its numbers.

Another unique crocodilian is the false gharial, or tomistoma, which, like the gharial, has a long, thin, rounded snout. It has fewer teeth than the gharial, and the snout is not quite so sharply demarcated from the rest of the head. Some features of the skull bones indicate that it is more closely related to the crocodiles than to the gharial, but it is distinct enough that most students of crocodilians give it a group of its own. It has several extinct relatives, including a huge saltwater species whose fossilized teeth are found in Florida phosphate deposits.[6] The false gharial is found in the southern part of the Malay peninsula, Sumatra, and the western part of Borneo. It seems to be an inhabitant of sluggish rivers and reaches a maximum length of a little under 16 feet. Little is known of its habits.

The remaining crocodilians are divided between the crocodiles and the alligators. External differences between the groups are few: the only reliable one is that in crocodiles the long fourth tooth of the lower jaw fits into a notch or opening in the side of the upper jaw, while in alligators the fourth tooth is not conspicuously longer than the others and the lower teeth fit into pits inside the upper tooth row. Crocodiles tend to have long, pointed snouts and alligators wide, blunt ones, but there are enough exceptions to make this rule unreliable. It is sometimes said that alligators open their mouth by moving the lower jaw, while crocodiles move the upper. This is not true; however, crocodiles may rest with the mouth open and the lower jaw pressed against the ground. To do this the animal tilts the entire top of its head upward. This gaping exposes the lining of the mouth to the air, causes cooling by evaporation, and helps control body temperature. Like many other reptiles, crocodilians maintain a fairly constant temperature by behavioral patterns such as basking. In the Nile crocodile, this is 25.6° C (78° F).[7]

Crocodiles

As with other giant reptiles, the maximum size reached by crocodiles is often a matter of whose record one chooses to accept. The preservation of these huge animals under field conditions is rarely possible, and even measurements taken in the field may be subject to considerable error. Estimates, particularly those made on living animals, can be fantastically inaccurate. The record accepted by many American herpetologists is one of 22 feet 4 inches reported by the German naturalist and explorer Alexander von Humboldt in 1800 for a specimen of the little-known Orinoco crocodile. The leading Old World contender for size honors is the saltwater crocodile. The skull of an alleged 33-foot specimen from Bengal is in the British Museum, and a famous man-eater killed on Luzon in the Philippines in 1820 was reported as measuring slightly over 29 feet. The skull of this crocodile is now in the Museum of Comparative Zoology at Harvard. Herpetologists who have recently measured these skulls say that if they came from specimens in the neighborhood of 30 feet long, the animals were microcephalic. A length of about 22 feet seems more reasonable.[8]

A British sea captain and trader named Alexander Hamilton, writing of his experiences in the East Indies about 1705, told how his men were disturbed by "alligators" as they were working on a stage rigged alongside their vessel. Musket balls fired at the saurians did them no damage until a lucky shot hit the largest in the eye. A day or two later a dead "alligator," presumably the same one they shot, was found on the shore nearby and was recorded as measuring 27.5 feet. This account undoubtedly refers to the saltwater crocodile, but the length may be an estimate.[9]

If there are 20-foot crocodiles left in the world today, their numbers must be few indeed. Paul Deraniyagala of Ceylon, who has combined outstanding careers in zoology, archaeology, and politics and is a talented artist as well, wrote in 1953 that Ceylon specimens of the saltwater crocodile 18 and 21 feet long had been taken "in recent years."[10] Wilfred T. Neill, an American herpetologist who has probably had field experience with more kinds of crocodilians than any zoologist now living, saw several 19-foot saltwater crocodiles in New Guinea presumably during World War II.[11] A 21-foot Nile crocodile was said to have been killed in 1949.

In addition to probably being the biggest of the crocodiles, the saltwater crocodile is of interest in being largely restricted to a brackish-water habitat. Although it seems quite at home in fresh water, it is seldom found far from the seacoast and is particularly common in estuaries, mangrove swamps, and the lower reaches of rivers. It occurs from the southern part of the Indian peninsula around the fringes of southeastern Asia to the Philippines and northern Australia. It makes long sea voyages and is regularly found on the Sunda Islands of Indonesia, the Solomons, and New Guinea. Exceptionally venturesome individuals have showed up on the Cocos-Keeling Islands, which are separated by more than 600 miles of empty ocean from the south coast of Sumatra, and on Rotuma Island, which is about the same distance from the Solomons, although the latter individual could have come by way of Fiji, which would have involved a mere 260-mile swim. Crocodiles are prominent in native legends in the Fiji Islands, but the resident population was apparently exterminated some time prior to 1900.[12] The saltwater crocodile shows no changes in body form adapting it to marine life, but its body armor is somewhat lighter than that of other big crocodiles.

Another large crocodile that inhabits salt or brackish water to a considerable extent is the American crocodile. Some recent writers credit it with a length of 20 to 23 feet, but the source of these records is unknown. Sixteen feet seems to be about the maximum length reached. Its distribution—on the major islands of the West Indies (except Puerto Rico), along the Atlantic coastal lowlands from Tamaulipas and the southern coasts of Florida to Venezuela, and on the Pacific from Ecuador to southern Sinaloa—is similar to that of the saltwater crocodile in Asia, although less extensive. Constantine Rafinesque, an eccentric early American naturalist, reported crocodiles in Florida in 1822, apparently on the basis of a shrewd hunch. It was not until 1869 that Dr. Jeffries Wyman of Boston obtained from Miami the first specimen to reach a scientific collection.[13] In January, 1875, William Hornaday, one of the founding fathers of the New York Zoological Society, shot a 15-foot male crocodile and a 10-foot 8-inch female in Arch Creek, which flows into Biscayne Bay in what is now North Miami. The male was a battle-scarred old veteran, but "In brightness of color, smoothness of armor, and litheness of contour the female greatly outranked her rough and burly lord."[14] Hornaday thought his crocodiles represented a new species rather than the *Crocodilus acutus* that had been described from Haiti in 1807.

At one time the American crocodile was found along the east coast of Florida to Palm Beach and along the west coast to Charlotte Harbor, with an isolated and somewhat dubious record for Lake Harney, slightly north of Cape Canaveral. Its range has gradually decreased until today the only known breeding populations are in Everglades National Park, on Key Largo, and on Big Pine Key. According to John C. Ogden, research biologist at Everglades National Park, six nests were located in the park in 1970 and seven in 1971, although two of the latter had been destroyed by raccoons. Florida is marginal habitat for the crocodile, both because of climatic conditions and because of competition with the better-adapted alligator.

Christopher Columbus's son, Don Ferdinand, saw "vast, great crocodiles" in Port Bastimentos on the coast of what is now Haiti in 1497 and said that, "they scatter a scent as if all the musk in the world were together; but they are so ravenous and cruel that if they find a man sleeping, they drag him to the water to devour him, though they are fearful and cowardly when attacked."[15] The American crocodile was locally plentiful in the Ocho Rios section of Jamaica until recently and occurs on Cuba with the endemic Cuban crocodile. Stragglers occasionally show up on the smaller West Indian islands.

On the mainland of tropical America, the American crocodile was formerly plentiful in many places and sometimes occurs well inland. The late Karl P. Schmidt, a distinguished zoologist formerly of the Field Museum in Chicago, described a veritable crocodile paradise at Lake Ticamaya in Honduras in 1923. In this warm, shallow lake, about 2 miles long, seventy-five crocodiles were counted in a single bay, and a sweep of the waters at night with a head lamp would show thirty to thirty-five pairs of eyes. As an example of the way large reptiles shrink at the approach of a tape measure, Schmidt said that "a 13-foot crocodile" killed by an American hunter the day he arrived at the lake actually measured 8 feet 10 inches. Surprisingly enough, the party saw no crocodiles under 3 feet long. The young must be very secretive in an effort to avoid their enemies, the most dangerous of these apparently being larger individuals of their own species. The hunters did find a crocodile nest, and the eggs provided a welcome change from the regular diet of rice and beans. The most tangible result of this expedition is the fine crocodile habitat group in the Field Museum. Harpooning and killing the crocodiles was only the beginning. The backbreaking work of making plaster molds of the freshly killed animals and skinning them before they decomposed in the 100-degree heat, the necessity of drinking foul-tasting lake water made even less potable by the addition of chlorine tablets, and the attacks of countless ticks and mosquitoes are aspects of scientific collecting seldom appreciated by those who see only the end result.[16]

Some recent experiments show that very young American crocodiles and saltwater crocodiles can tolerate full-strength sea water only 4 to 6 days, while somewhat larger animals weighing about 7.5 pounds are not seriously troubled by 5 months in this element.[17] The manner in which they adapt to salt water is not known. It is possible that salt-excreting glands similar to those found in sea turtles, sea snakes, and the marine iguana become activated, although no direct evidence of this was found. The "crocodile tears" so often mentioned in folklore may be, like the tears of sea turtles, a way in which the animal excretes excess salt, and it may be significant that tear shedding has not been reported for alligators and caimans which live almost exclusively in fresh water.

If man evolved from his prehuman primate ancestors along the lakes and streams of eastern or central Africa, he did so under the hungry eyes of the Nile crocodile. Probably no crocodilian has been so long or so intimately involved with the human species. It saw the first great civilization grow along the valley of the Nile and was the first of its tribe known to the peoples of Europe. It may once have inhabited practically all of Africa, including areas that are now desert. A small, isolated population still lives in rocky pools in the guelta of Archei, a deep, narrow canyon in the southern Sahara far from any other permanent lake or stream. The natives there regard the crocodiles as beneficial scavengers that help in keeping the pools clean.[18] Crocodiles once lived around the periphery of the Mediterranean from Tunisia to Israel and may have occasionally wandered to Sicily. The giant snake killed by the Roman troops of Attilius Regulus near the River Bagrada during the third Punic War (ca. 149 B.C.) may have actually been a crocodile. The creature was said to have been 120 feet long and to have devoured many soldiers before it was finally dispatched with heavy siege weapons.[19] The Bagrada (now known as the Medjera) is a comparatively small stream flowing into the Mediterranean near the modern city of Tunis. This is far from any location where there are records of the African python, the only snake that might have inspired such an account.

In 1525, a French physician on a pilgrimage to the Holy Land was allegedly carried off by a crocodile while bathing in the Jordan River.[20] Crocodiles survived in the Kabara swamps south of Haifa, Israel, until about 1900. Although less inclined to enter salt water than some other large crocodiles, the Nile crocodile inhabited the Comoro and Seychelles Islands until about 1820 and is still found in Madagascar. However, it may have reached these islands when they had closer connections with the African continent.

Major Stanley S. Flower, a British zoologist and soldier with much experience in North Africa and the Middle East, said that crocodiles were virtually exterminated from the Egyptian portion of the Nile at about the turn of the century, although they remained common in the Sudan. They returned to lower Egypt in fair numbers during the time of World War I and immediately thereafter. One 12 feet long was killed on the Rosetta branch of the Nile about 80 miles north of Cairo in January 1927.[21] If crocodiles occur in Egypt today, they must be rare stragglers from the south. There were no specimens of the crocodile among the 3,424 reptiles and amphibians collected in Egypt by U. S. Naval Medical Research Unit No. 3 between 1949 and 1965.[22] Early accounts give the length of the Nile crocodile as 18 cubits (about 27 feet), and *Brehm's Tierleben*, a standard European work on natural history in the early part of the twentieth century, says it may reach a length of 10 meters (about 32.5 feet), but these are unsupported estimates.

Flower said that the largest crocodiles he encountered were in the White Nile during the early part of this century. He gave measurements of six that were 14 feet or longer. The largest measured 16 feet 7 inches, but he estimated that others were larger and may have approached 20 feet.[23] Hugh Cott, a British zoologist whose work on adaptive coloration in animals led to a wartime career as a camouflage specialist, carried out and published a field study on the Nile crocodile in East Africa that is undoubtedly the best work to date on the ecology and economic importance of a crocodilian. Although the largest crocodile he personally examined measured a little over 15 feet, he assembled what he considers reliable records of considerably larger individuals. The biggest ones were taken in Emin Pasha Gulf of Lake Victoria, where, in 1905, when this was German territory, the Duke of Mecklenberg shot a crocodile 6.5 meters (about 21.5 feet) long. Erich Nowotny, a professional hide hunter, reported killing a 21-foot crocodile there in 1949 and seeing another that he estimated as even larger. Cott believed that crocodiles this size must be well over 100 years old, but this may be too generous an estimate.[24] Given the right genetic combination, good health, and a rich food supply at the right time, reptiles, including the giant types, are capable of amazingly rapid growth and can reach sizes well above the average for the species within a normal life span, which, in the case of most crocodilians, seems to be 25 to 50 years. At the other end of the size spectrum, some populations of the Nile crocodile, such as those of the guelta of Archei, are stunted and do not reach a length of much more than 6 feet. This may result from natural selection for smaller size plus a restricted food supply.

Asia is the home of another well-known large crocodile, the Indian crocodile, or mugger. The latter name is a corruption of *magarmach*, an Urdu or Hindi word for a crocodile of any kind. The species is also often called marsh crocodile, a translation of its scientific name, *palustris*, though it seems to inhabit lakes and sluggish streams more often than marshes. The center of its range is the Indian subcontinent: it occurs from the Dasht River on the Iran-Pakistan border to Assam and Burma, with a subspecies on Ceylon. This is the most snub-nosed of the big crocodiles, but the snout of an 18-inch preserved specimen in our collection is still noticeably longer and more pointed than that of an American alligator of about the same length. Thirteen feet is the maximum length reliably attributed to the Indian crocodile; however, it is proportionally heavy-bodied.

Many of our recollections of this crocodile center about a little river that flows into the sea a few miles west of Karachi. Alexander the Great may have known this river as the Arabius; today it is known as the Hab, and it forms the boundary between lower Sind and the Baluchistan state of Las Bela. Late in February 1960, Walter Fairservis, an archaeologist associated with the American Museum of Natural History, and I set out to trace the Hab as far toward its source as we could. For nearly a hundred miles we followed its valley northward through an almost uninhabited land. In the open areas, we flushed numbers of sand grouse and gray partridge. Chunky dust-colored spinytail lizards watched from the mouths of their burrows, and an occasional gazelle loped across the camel track that served as a road. From either side the precipitous, barren mountains slowly converged upon us, and the track grew rougher. Finally we reached a point beyond which not even a jeep could go. Not far from there a large clear stream fringed with trees flowed from the mountains into the river. Anhingas, moorhens, and a little pied woodpecker, birds totally foreign to the desert, suddenly appeared. It was a unique, unspoiled spot and an irresistible campsite. Soon after dark I swept the beam of my head lamp across the sheet of water in front of the tent. At the edge of the far bank, two eyes glowed red as coals. "A crocodile, Walter! It's got to be a crocodile! I've shined alligators in Florida, and this looks exactly the same."

Confirmation came next morning. We walked down the stream to a point where it was bordered by cliffs rising 50 feet or so above the water. Suddenly Walter grabbed my shoulder and pointed to the pool below. At the surface, like a child's toy in a bathtub, floated a crocodile about 5 or 6 feet long. We moved a little, and a flick of the creature's tail sent him into hiding in the deep water below the undercut stream bank.

Madge and I with our three girls visited this spot about a month later, accompanied by Major Robert Brewer, a U. S. military adviser, and his family. Again we saw crocodile eyes by night, but none of the animals showed themselves in the daytime. The major got a shot at one pair of eyes but with no observable results. It was on this foray that our eldest daughter slipped off the bank into a deep pool. We hauled her out shaking with indignation and fright. Although I had assured her that these crocodiles were not man-eaters, she was by no means convinced, and moreover she thought that being shot at might be enough to make a normally peaceful crocodile hostile. On another trip, Madge and I found another fine crocodile pool in the main stream of the Hab. On one side, where there was a low rock cliff, the water was about 10 feet deep; on the other, gravelly shore, it was shallow. Here too the crocodiles floated near the shelter of the cliff by day and came into the shallows at night. Our daughter April surprised a small one on the bank about sunrise, but it quickly slid into the water.

Alligators and Caimans

When the Spaniards began to explore what is now the southeastern United States, they encountered a large reptile which they referred to simply as "el lagarto" (the lizard), but which they evidently recognized as distinct from "el cocodrilo," the crocodile of Africa and Asia. Soon corrupted to "aligarto" and eventually to "alligator," this name has been applied to crocodilians in many parts of the world. In Cuba, southern Asia, Australia, and perhaps in other places where two obviously different kinds of crocodilians are found, Europeans and Americans tend to call one crocodile and the other alligator without much regard for the zoological meaning of these terms. Strictly speaking, there are only two species of alligators, one in the southeastern United States and the other in the lower Yangtze valley in eastern China. The presence of closely related reptile species in eastern Asia and the eastern United States is not surprising. It occurs with rat snakes of the genus *Elaphe*, pit vipers of the genus *Agkistrodon*, and lizards of the genus *Eumeces,* to name but three. In most of these cases, study of living species and of the few fossils that exist indicates that the group originated in Asia and migrated to America. For the alligators and their recent and fossil relatives, however, the evidence is more difficult to interpret. Although fossils of alligatorlike reptiles are known from the Eocene and Miocene epochs in Europe, the earliest fossils are American, and the group seems to have undergone most of its evolution and diversification in the Americas.

The Chinese alligator shared its land with an old and contemplative human society; the American alligator with a young and aggressive one. For both animals this meant a drastic reduction in range and numbers, taking place over a span of centuries in China and largely a matter of the last 150 years in the United States. However, the difference in human temperaments led to the American alligator's being the best known of crocodilians scientifically, while the Chinese alligator has remained largely in the realm of legend and folklore.

It is often said that the imperial dragon of China is based on a crocodilian of some sort, possibly the alligator, which is mentioned in Chinese literature dating to at least the Wu Dynasty (A.D. 222–227). The extent of Chinese culture in southeastern Asia is so great, however, that the Chinese must have become acquainted quite early with some of the larger and more formidable crocodilians, such as the saltwater crocodile, which was reported near Hong Kong as recently as 1912.

Marco Polo evidently knew of some alligatorlike animal, although, like most early writers, he considered it a sort of snake. On his travels in the province of Carajan (a part of Yunnan), he wrote:

> Very great serpents are bred in this country, some of which are ten paces in length and in thickness ten spans. They have two little feet before near the head, with three talons or claws like lions, and the eyes are bigger than a loaf, shining very bright. They have their mouths and jaws so very wide that they are able to swallow a man, great and sharp teeth; nor is there any man or other living creature which can behold these serpents without terror. There are also some less, of eight, six, or five paces long, which are taken after this manner: in the daytime they use to lie hid, by reason of the heat, in holes, out of which they go at night to seek their prey and devour whatsoever they get, lions and wolves, as well as other beasts, and then they go to seek water, leaving such a track through their weight in the sands as if a piece of timber had been drawn there; whereupon the hunters fasten under the sands great iron spikes, in their usual tracks, whereby they are wounded and slain. The crows presently proclaim the serpent's fate, and by their cries invite the hunters who come and flay him, taking out his gall which is used for divers medicines, amongst other things for the biting of mad dogs, a pennyweight given in wine; and for women in travail; for carbuncles and other distempers, and they sell the flesh dear as being exceeding delicate.[25]

Henry Hamel, secretary or yeoman of a Dutch ship wrecked on the coast of Korea in 1593, wrote:

> In Korea the rivers are often pestered with alligators or crocodiles of several sizes; some are 18 to 20 ells long; the eye is small but very sharp; the teeth placed like those of a comb. When they eat they move only the upper jaw . . . three children were once found in a crocodile's belly.[26]

Yunnan and Korea are both far beyond the range of any crocodilian as known to nineteenth- and twentieth-century science. Probably both Marco Polo and Hamel got their information second or third hand, and it is of little help to the zoogeographer, although it suggests that the range of the alligator in China was formerly more extensive.

It was scarcely a century ago that Western science became aware of the Chinese alligator. The first zoologist to write of it was the Englishman Robert Swinhoe, whose account reads in part:

> In February 1869, some Chinese were exhibiting in the native city of Shanghi [*sic*] what they called a dragon, which they declared had been dug out of a hole in the province of Shense. It was a young crocodile about four feet long. . . . They made so much money by showing it that they refused to sell it. I cannot, of course, guess its species.[27]

It was recognized as an alligator rather than a crocodile by A. A. Fauvel of the Imperial Chinese Maritime Customs, Honorary Curator of the Shanghai Museum. He officially described it and published a historical account of it in 1879. At that time the reptiles were sometimes kept in temples near Nanking.[28]

Clifford Pope, an American herpetologist who collected extensively in China between 1921 and 1926, found the Chinese alligator along the Chingshui River near Wuhu, which is not far from Nanking. Here the alligators lived along the river in plains that were uninhabitable because of frequent floods. These plains were covered with high grass during much of the year and were frequented by great numbers of water birds and small wild animals. Of the capture of the alligators he wrote:

> Being strongly advised that the presence of a foreigner would make prices soar, I remained in the city and on March 14 sent out two scouts. Late on the nineteenth one of the scouts proudly strode in ahead of a rickshaw loaded with five securely bound alligators. Their sluggishness showed that they had been roused from hibernation; the most active made feeble movements and roared a little.
>
> Less than half a mile from a large village the specimens had been dug out of burrows, each a foot in diamter and about five deep. A wildcat that dashed out of one was chased until exhausted and caught. . . . The ground was so bare that not even a cat could find a hiding place.[29]

In the summer when the plains were inundated, no alligators could be found. Pope's total outlay for a series of nineteen specimens was $19.60. If a Chinese alligator were to become available on the American market today, the price would start at about $1000.

The Chinese alligator is one of the smaller crocodilians, attaining a maximum length of 5 to 6 feet as based on specimens taken during the last century. If accounts from Chinese literature can be believed, it once reached a considerably larger size.

At the time of Pope's visit to China, the alligator was known from four provinces, all in the lower Yangtze drainage, but occurred only in small colonies believed to be remnants of a once extensive range. Events in China since then can scarcely have been favorable for the alligator. Some conservationists believe it is already extinct as a wild animal; however, the government of mainland China seems to have an enlightened policy as regards endangered wildlife and may have legislation to protect some of the surviving colonies. It is to be hoped that herpetologists may follow the table-tennis players, and that the next few years will bring some information as regards the status of what may be the rarest of crocodilians.

The two species of alligators are the only crocodilians whose ranges are wholly confined to the temperate zone. At the time of the settlement of the southeastern United States, the American alligator was extremely abundant in places. William Bartram, an eighteenth-century naturalist and artist who explored extensively in Florida with his father, a pioneer American botanist, wrote in 1791 that along the St. Johns River "it would have been easy to have walked across on their heads had the animals been harmless."[30]

The northernmost limit of the alligator's range in the United States seems to have been the southern shore of Albemarle Sound, North Carolina, where it gave its name to the Alligator River. In 1947 and 1948, when my brother and I were collecting reptiles in the Mississippi valley, we began asking the country people in the boot-heel of Missouri about alligators. We got only negative or facetious answers there and farther south until we came to a place called Greenfield Bayou, near Jonestown, Mississippi, about 40 miles southwest of Memphis, Tennessee. Here the people said there had once been plenty of alligators. Although there were now uncommon and very shy, a large one had been killed in a cotton field in the summer of 1947. Up to 1953, alligators were occasionally found along the Red River, which forms the Texas–Oklahoma border, as far as Denison Dam, although they seem to have been occasional stray individuals rather than a breeding population. There are also reports of alligators from the Rio Grande valley near Brownsville and Eagle Pass.

The American alligator stands up well in the size competition with other crocodilians. The late E. A. McIlhenny, an amateur naturalist and member of an old Louisiana land-owning family, grew up with alligators as a part of his everyday life on the family plantation at Avery Island in the southwestern part of the state. As a child he and his companions had names for all the bigger alligators that lived in the bayou near his home. The largest, "Old Monsurat," was killed by the boys' tutor in the fall of 1879. This alligator measured 18 feet 3 inches. It took four mules to haul the monster to the house. McIlhenny mentions several other alligators 17 to 18 feet long, the last one killed in 1916. The giant of all was one that McIlhenny himself shot on January 2, 1890, when it was lying chilled and torpid in a marsh. The alligator was so large it could not be dragged onto solid ground for skinning. An average of three measurements determined its length as 19 feet 2 inches. This was a very old alligator with its teeth worn almost to the jawbone. McIlhenny said he believed the reptile had grown too feeble to hibernate and would probably not have survived the winter.[31] The largest Florida alligator measured by Wilfred T. Neill was one 17 feet 5 inches killed in 1956 at Lake Apopka. It was a healthy, vigorous animal showing no sign of senility and might well have grown another foot or two had its life not been cut short.[32] "Big George," a Florida alligator exhibited for many years at Ross Allen's Reptile Institute at Silver Springs, Florida, measured 14 feet 9 inches when he died in 1967 at an estimated age of 59 years. His increase in length during his last 17 years in captivity was almost nil. Today an 8-foot alligator is large and one of 12 feet is near record size. The giants of 15 feet and more will probably never be seen again, but if they are it may be as a result of the twentieth century's most dramatic advance in technology. We were recently told that near an atomic energy plant on the Savannah River alligators are increasing in both size and abundance. Hot water discharged into the stream permits them to be active the year around and has brought about an increase in the fish and turtles on which they feed, while tight security regulations protect them from hunters.

In its broader sense, the alligator group includes the caimans of South and Central America. This name seems to be of Carib Indian origin but was soon assimilated into Spanish and Portuguese and applied to crocodilians generally. Kayman's Gate in the city of Colombo, Ceylon, got its name during the Portuguese occupation of that island. In Venezuela and Colombia today, the two species of true crocodiles are usually called caimans, while the reptiles that herpetologists know as caimans are known as *babas* or *babillas*. Jacare or Yacare, a Guarini Indian word meaning "floating log," is also often applied to caimans in South America.

The caimans differ from the American and Chinese alligators in lacking a bony septum between the nostrils. In addition, the bony plates, or osteoderms, that make up the "hornback," or dorsal armor, of all crocodilians are found in caimans under the belly skin. In alligators these ventral osteoderms are absent or weakly developed.

In addition to having American affinities, the alligators and caimans are more addicted to fresh water than are the crocodiles and show little tendency to spread along coastlines or to reach offshore islands. Instead their distribution is correlated with river drainage systems.

The five species of caimans are found from the Isthmus of Tehuantepec in southern Mexico to northern Argentina. The black caiman of the Amazon reaches a length of 15 feet and is considerably the largest member of the group; none of the others exceed 8 feet.[33] The spectacled caiman, so called because of the curved bony ridge across the snout in front of the eyes, and the closely related brown caiman were abundant until recently in Central America and northern South America. About 1945, when laws prohibiting the sale of young alligators went into effect in Florida, baby caimans largely replaced them in the pet trade, and many thousands were sold. Most survived only a short time, for the caiman adapts even less well than the alligator to living in a fish bowl and being offered a diet of dried flies, ant eggs, and the other rubbish sold as food for such creatures. Nevertheless, escaped or liberated pet caimans have been found in the field in several southern states, and some herpetologists believe the species may be establishing itself in southern Florida. One of the sillier folktales of the late 1960s was that the New York sewers were becoming infested with alligators, presumably unwanted pets that had been flushed down the toilet. In some of the accounts, these were growing to formidable size from feeding on rats. We have been unsuccessful in tracing the source of these legends but would assure New Yorkers that alligators are not among their urban problems. There is, however, a well-documented account of a released pet alligator that survived in a Pennsylvania marsh for 6 or 7 years and grew to a length of about 5 feet before it was killed.[34]

The dwarf and smooth-fronted caimans of northern South America rarely exceed 4 feet in length. They are curious little dragons with long, recurved teeth and the heaviest and most complete dermal armor of any crocodilians. This may serve to protect them from injury when swept against rocks in the rather swift streams they inhabit. They are also reported not to show "diving bradycardia," a slowing of the heartbeat on submergence that is seen in many diving reptiles, birds, and mammals, including, to some degree, experienced skin divers.[35] If the laboratory specimens tested for this were not abnormal, it may mean that these caimans are less efficient divers than other crocodilians.

3. Crocodilian Life Patterns

Breeding and Nesting

From the human viewpoint, the sexes in crocodilians are difficult to distinguish. The reptiles themselves seem to have no difficulty on this score; the cues they use are evidently chemical, auditory, and behavioral. In most, if not all, crocodilians the male is larger than the female. Average male adults of the American alligator today are 7 to 9 feet long and females 5 to 7 feet.[1] For the Nile crocodile in East Africa, males mature at a little more than 9 feet and often reach 12 to 13 feet; females mature at slightly over 8 feet and seldom exceed 11 feet.[2]

Crocodilians of both sexes have musk glands opening close to the vent. Reasoning mostly by analogy with snakes and lizards, researchers think that these secrete a substance that aids the sexes in finding each other. Females of the Nile crocodile and American alligator are reported to lay a trail in the water that the male follows. McIlhenny tells of seeing a line of eight large male alligators following a female at intervals of about 50 feet and roaring periodically.[3] All crocodilians also have a pair of eversible tubular sacs opening near the edge of the lower jaw. Some peoples of southeastern Asia believe that these are an cxtra pair of eyes used by the reptiles for underwater vision; their actual function is not well understood. They contain a waxy substance that many authorities say has a pungent, musky odor, although some report it practically odorless. The sacs are everted under conditions of stress, such as defense of the nest by females, attacks by predators, or battles with large prey. They seem to be quite as functional in small juveniles as in adults and have no known association with sexual activity.

The courtship of the American alligator takes place in spring and lasts several days. During this time, the male strokes the female's back and rump with his forelimbs, rubs her throat with his head, and blows streams of bubbles past her cheeks.[4] Female Nile crocodiles have been seen to rear high out of the water before the male, with jaws wide open and pointed to the sky.[5] A similar display has been seen with the Indian crocodile.[6] During courtship the male and female of the American crocodile raise their heads vertically above water and rub each other's heads and necks. At the same time they make grunting or snoring noises and beat the water with their tails.[7]

The actual mating of crocodilians has rarely been observed even in captive animals. Many early writers assumed that it took place on land, and some added some egregiously fanciful touches:

> The female, which in the act of copulation is laid on her back, has much difficulty in rising again; it is even said that she can neither change her position nor turn without the assistance of the male. Will it be believed that in Upper Egypt there are men to be found who, hurried on by the excess of an unexampled depravation and bestiality, take advantage of the forced situation of the female crocodile, hunt away the male and replace him in a commerce that shocks humanity? Horrible embraces, frightful enjoyments, the knowledge of which should never have stained the disgusting pages of the history of human perversity![8]

Recent observers of crocodilian matings are agreed that they usually occur in shallow water. Male lizards of many species hold the female with their jaws during copulation, but this is not the custom of crocodilians, although there are some reports of male crocodiles' biting the female about the neck during courtship. Instead, the male grasps the female more or less gently around the neck with his forelimbs and bends his body under hers. She, in turn, may lie partly on her side so that her vent is turned to receive the penis. Coitus may last from a few minutes to several hours.[9]

Crocodilians have two basic nesting patterns. The gharial and most of the large crocodiles excavate a hole in sand or gravel and deposit their eggs in two or more layers, finally covering them and smoothing the soil over them. The American alligator, saltwater crocodile, African sharp-nosed crocodile, tomistoma, and at least some of the caimans make a nest mound of vegetation and soil, within which they lay their eggs.[10] McIlhenny reported in detail on alligator nesting in Louisiana. It begins with the female's making a clearing near the water by mashing down and biting off the vegetation. Trees up to 3 inches in diameter may be broken off. She then gathers in her mouth a considerable quantity of leaves, twigs, rushes, and other vegetable debris, which she shapes into a conical mound about 3 feet high. The material in the upper part of the mound is wetter than that at the base. The female hollows out the center of the mound with her hind feet and lays her eggs. Then she covers the eggs and compacts the top of the mound by crawling over it repeatedly. In one instance, the entire process took most of three days.[11] Neill's account, based on experience in Georgia and Florida, differs in several important respects. There is no preliminary clearing of the nest site, and the female was never observed to use her mouth to carry nest material. The material is scraped up and the nest shaped by movements of the female's body and tail.[12] If a nest site is not seriously disturbed, a female may use the same one for several years, adding somewhat to it each year.

The behavior of other nest-building crocodilians has not been reported in detail. In Ceylon, the distribution of the saltwater crocodile is closely associated with that of a flag plant that it uses in its nest mounds.[13] The American crocodile is interesting in being a species apparently in transition between the two nesting patterns. The female buries her eggs in comparatively high but moist ground. During the incubation period she visits her nest regularly and may scrape additional soil onto it with her hind feet until a low mound is formed.[14]

Communal nesting is reported for some crocodilians. At about the turn of the century, the Swiss naturalist Gottfried Hagmann observed and photographed enormous numbers of caimans on Mexiana and Marajo islands near the mouth of the Amazon. Here the spectacled caimans assembled for nesting in May and June, and the black caimans in October and November.[15] Cott described communal nesting grounds of the Nile crocodile at Lake Albert in Uganda, one of the few places where the animals had not at that time been subjected to intense hunting.[16] In Florida, about 1765, Bartram saw a great number of alligator nests, "resembling hay cocks, ranged like an encampment along the banks."[17] Whatever advantages the communal nesting habit had seem to have been nullified by the advent of human hunters, and it apparently has been abandoned by the reptiles over most of their ranges.

The clutches of most of the larger crocodilians contain 20 to 70 eggs, which are oval and have a hard, brittle shell. As is a general rule with reptiles, larger females lay larger clutches. For most species, the size of the egg is between that of a large hen's egg and a goose egg. In the few species for which it is known, the incubation period under natural conditions is about 60 to 90 days.

With all the well-known species, the female remains near the nest. As with many other reptilian traits, the degree of attention she gives to the eggs varies with the species and individual, as well as with interpretations of her actions by human observers. The Nile crocodile seems to be unique in leaving the water and spending most of the incubation period lying directly on or near her nest.[18] The saltwater crocodile, which also attends her nest closely, is reported to dig a trench from the water to her nest and lie in it.[19] One account says she keeps her nest mound moist by splashing water onto it with her tail; this has been questioned by some authorities. Most of the other species seem to lie in the water close to the nest mound and visit it occasionally. Female alligators help keep their nests damp by crawling out of the water onto them. With the Nile crocodile, the guarding of the nest is largely a passive process, the mere presence of the female being enough to deter most potential egg thieves. In Africa, the worst nest raider is the Nile monitor, a large lizard with an insatiable appetite for eggs. The lizards sometimes work in pairs, one distracting the female while the other digs up the eggs. Mongooses, baboons, and hyenas are other important egg predators, while the marabou stork frequently eats eggs from nests that have been uncovered by other raiders.[20] In the Americas, peccaries, bears, raccoons, skunks, and otters are among the creatures that relish a meal of crocodilian eggs.

A most important role of the attending female is that of liberating the newly hatched young from the nest. In Uganda and Northern Rhodesia, where Cott made his observations, and presumably over much of the rest of Africa and the Indian subcontinent, crocodiles nest during the dry season when water levels are low. At the end of the incubation period, the soil over the nest may be baked to an almost bricklike hardness, and it would be almost impossible for the newly hatched young to escape. The hatchlings signal their mother by a yapping or quacking call that apparently can be given while the reptile is still within the egg. The Nile crocodile female unearths her young by rotary, scooping movements of her body that produce a characteristic broad shallow crater; she does not dig with her feet.[21] The techniques used by other crocodilians have not been reported. It is when the young are ready to emerge that the brooding female becomes most aggressive in defense of her nest and may attack any intruder, including man. In the case of several species, this protection extends to the post-hatching period; the young have been seen clustered about the mother or even resting on her back. In the Okefenokee Swamp on June 29, 1971, we saw three young alligators about 15 or 16 inches long resting on waterlily pads. When our guides imitated the high-pitched "umph-umph" call of a baby alligator in distress, a female about 5 feet long promptly popped her head out of the water, hissing and snorting, her jaws partly open. More annoyance caused her to lunge at the boat and snap her jaws. While she may have also had a nest nearby, there seems no doubt that she responded in defense of young which must have been nearly a year old at the time. We went on our way with the feeling that this particular alligator was a devoted and plucky mother. Neill had a similar encounter with a New Guinea crocodile, when the distress cry of a newly caught baby brought an angry adult charging to the rescue.[22] It is hard to say how resolutely such an attack would be pressed home against a human intruder. There seem to be no reports of human fatalities or serious injuries unquestionably caused by defending female crocodiles or alligators, although in many instances it is hard to say what motivated the reptile to attack. The most dangerous attacks seem to be made by hungry crocodiles or by large males defending their territory.

The bellowing or roaring of the American alligator has often been considered an activity of males during the breeding season; however, both sexes indulge in this, the males being noisier because of their greater size, and most of the bellowing occurs during summer when the mating season is over. The sound is a single, deep, vibratory note repeated five to eight times. It has been likened by various writers to the bellow of a bull, distant thunder, the roar of a lion, or the roll of a bass drum. Noises such as thunder, blasting, and truck engines sometimes stimulate alligators to bellow in reply.[23] There is no clear proof that the roar attracts females or repels rivals, nor is it accompanied by aggressive behavior. It simply seems to be an alligator's way of saying, "I am here!" The American alligator is the loudest of the crocodilian clan, but similar vocalization has been reported for caimans and some crocodiles.

Food and Feeding Habits

Like nearly everything else about large reptiles, the voracity of crocodiles has often been exaggerated. The adults particularly lead a lazy life and seem to get by on about one full meal a week. Smaller animals eat proportionally more, but their food intake is about 0.36 to 0.28 kilograms (0.8 to 0.6 pounds) of food daily as compared to 2.8 kilograms (6 pounds) for a pelican of about the same weight.[24] Although tens of thousands of crocodilians are killed and skinned every year, few of the skinners have bothered to examine and record the food in the stomachs of their victims. Those who do so are much more inclined to report the macabre or bizarre than the regular food of the animals. The finding of a diamond ring and twenty gold sovereigns in one stomach is certainly noteworthy but tells the biologist very little; the stomach of a South African crocodile that contained twenty-two dog license disks proves only that this animal was remarkably successful at catching dogs, and licensed ones at that.[25] To get a true idea of what crocodilians eat, it is necessary to go to one of the few scientific studies, such as that of Cott on the Nile Crocodile. He found that insects (chiefly aquatic beetles and giant waterbugs), spiders, and frogs were the principal foods of crocodiles under 1.5 meters (about 5 feet) long. The largest crocodiles, 10 feet or more, fed chiefly on mammals, such as the various species of antelopes, young hippopotamus, domestic cattle, sheep, and goats, and on reptiles, a high percentage of which were smaller crocodiles. Fish were eaten by crocodiles of all sizes except the smallest young and were the principal food of intermediate-sized reptiles (2 to 3.5 meters or about 6 to 10.5 feet). In only one of the eight localities sampled was the predominant fish species eaten a commercially valuable one. Birds, particularly diving species such as cormorants and anhingas, were eaten by all but the smallest specimens but were never a major food item. A surprising finding was the high prevalence of snails in stomachs of crocodiles of all sizes. In some localities they made up more than 80 percent of the food items.[26]

The food of the other well-studied crocodilian, the American alligator, is basically much the same as that of the Nile crocodile. The food of the young reptiles consists predominantly of aquatic insects, water snails, and crustaceans such as crawfish and shrimp, with fish, snakes, and frogs making up the balance.[27] Larger alligators eat many fish and turtles. Rough fish such as bowfin, catfish, and gar (except the longnose gar) are most often eaten; largemouth bass and other game fish are seldom eaten except when found dead or dying in oxygen-depleted pools. The large turtles known through much of the south as "cooters" are a favorite food of the alligator, as are the softshelled turtles. The high, heavy shells of the Florida red-bellied turtle and the peninsula cooter may give them some degree of protection, for they tend to be more abundant than other turtles in places where alligators are common. Snakes are eaten. The big, heavy, but nonpoisonous watersnakes are unceremoniously gulped down; the superficially similar but venomous cottonmouth is crushed and shaken to death before being swallowed. Muskrats, rabbits, raccoons, and other medium-sized mammals, including dogs and young pigs, are eaten with some frequency. Waterfowl are eaten regularly but seldom in large numbers. The very largest alligators may take larger mammals such as deer, adult pigs, and fairly well grown cattle.[28]

The special food preferences of other crocodilians are little known. The gharial and other species with long narrow snouts are assumed to be almost exclusively fish-eaters. Captive gharials feed by lying quietly on the bottom of their tank and catching fish by a sudden sidewise sweep of the long thin jaws. Even fast-swimming fish such as shiners are readily caught.[29] Gharials are also known to eat dogs and waterfowl occasionally, and there is a report of one that had managed to down the hindquarters of a donkey,[30] which would certainly be a most exceptional occurrence. The broad-snouted and Paraguay caimans feed heavily on water snails and have been ascribed an important role in controlling infections by the liver fluke, *Fasciola hepatica,* which is a serious parasite of sheep and cattle and occasionally affects man. This worm, like others of its kind, has an intermediate larval stage in a snail, so animals that reduce the snail population make it difficult for the fluke to complete its life cycle. With the killing of much of the caiman population in the Paraná River, there has been an increase in the number of snails and a corresponding increase in liver-fluke infection.[31]

Fish are important in the diet of the black caiman and the spectacled caiman. However, the killing off of most of these reptiles in the central Amazon basin has not led to better fishing, but rather the contrary—a decline in the fish population has paralleled the disappearance of the caimans. The reason for this paradox lies in the complicated relationships among living things in tropical river systems. The larger fish that live in the warm, nutritionally rich waters of the Amazon and its main tributaries spawn in enormous numbers in fluctuating lakes that are actually shallow expansions of smaller tributary streams. The water flowing into these lakes is very poor in certain minerals and organic nutrients essential for the aquatic plant life that supports water insects, crustaceans, and other invertebrates. These plants and invertebrates are the food of the newly hatched fish. Higher in the food pyramid are predatory fish, turtles, otters, and birds. At the top are the caimans. The life of the lake depends upon the enrichment of its waters by the excreta of these larger animals. When the caimans were numerous, they flocked into these lakes to feed on the spawning fish and the other animals that were also attracted to the area. As they fed, they also acted as so many amphibious manure spreaders. When they were removed, the lakes lost an important source of nutrients, the plant and invertebrate populations declined, and most of the fish fry starved.[32] This cycle is but one of many that sustain the rich and varied life of the wet tropics. Although most people still think of these regions as prodigally and inexhaustibly productive, they are actually an exquisitely fragile and delicately balanced environment. Attempts to exploit their resources or divert the land to other uses are likely to be disastrous unless carefully planned.

In attacking prey at the water's edge, the crocodilian quietly submerges well offshore and rapidly swims to the spot where its prey is. It then captures the animal in its jaws with a sudden upward rush. It may also use this method to catch flying birds or insects or leaping fish. Large Nile crocodiles are reported to hunt ashore, lying in ambush along trails used by animals coming to the water to drink. The prey is seized and usually dragged to the water, although it may be eaten on land.[33]

A remarkable series of photographs recently taken in Kruger National Park, South Africa, show a female impala attacked by a crocodile, which is promptly driven away by a hippopotamus. The hippo then nudges the injured antelope out of the water and treats it with what a human observer would interpret as a mixture of interest and solicitude, but it soon dies of its wounds. As the hippopotamus walks away, circling vultures descend, but in the final scene the patient crocodile crawls from the water to reclaim its prey.[34] It would be easy to fit a folk tale or fable to an incident such as this.

Fish or other aquatic prey may be actively pursued by crocodilians but are more often captured from underwater ambush.

Large fish may be battered against rocks or dragged ashore. Large prey is swallowed with the reptile's head out of water. If the animal is too large to be swallowed whole, crocodilians resort to a technique also used by some lizards. Fixing its teeth firmly in the flesh of the prey, the reptile rotates on the long axis of its body until a piece is torn off. This is then tossed or shaken about until it is in proper position for swallowing. I have seen captive alligators and crocodiles feed on animal carcasses in this manner. It is highly effective but not particularly pleasant to watch.

A still controversial point is the extent to which crocodilians keep "larders" in which freshly killed prey is stored until decomposition makes it more easily eaten. Neill found the bodies of young wild pigs that had apparently been buried in mounds of dried grass by the New Guinea crocodile. A crocodile about 5 feet long (adult size for this small species) was lying near one of the mounds.[35] The Indian crocodile has been reported to bury fish.[36] Cott found little evidence that the Nile crocodile has this habit, despite popular belief to the contrary. However, Paul L. Potous, a South African professional crocodile hunter, tells of an African native who was seized by a crocodile as he got out of a canoe and was dragged under water until he was almost drowned. He regained consciousness in a small cavern formed by partial collapse of the bank; decomposing remains of animals were in the cavern. The man escaped through a hole in the roof. When he reached his village after dark, he found he had already been given up as dead and had great difficulty convincing his kinsmen he was not a malignant ghost.[37]

The presence of stones in the stomachs of crocodilians has been observed since very early times and is often a part of the folklore surrounding these reptiles. According to one Malay legend, the crocodile swallows a stone whenever it enters a new bay or river. In another legend, it keeps tally of its human victims in this manner.[38] There seems no doubt that the reptiles swallow the stones deliberately and go to considerable trouble to find them. Nile crocodiles living in virtually stoneless swamps take longer to acquire a load of stones equal to that of animals living in rocky streams, but eventually they do so.[39] If stones are very hard to find, as they are in many swamps inhabited by alligators, the animals may swallow pine knots, soft-drink bottles, and pieces of metal, such as the bases of shotgun shells. Glass and metal bangles found in stomachs of crocodiles in Africa and India have sometimes unjustifiably been considered evidence of man-eating. It is much more likely that they were found on the bottom of streams and swallowed because they were hard objects. The Paraguay caiman may not have the habit of swallowing stones, for none were found in the stomachs of thirteen specimens examined.[40] Perhaps these animals were taken in a habitat where stones or similar objects were especially hard to find, or possibly this species has found an anatomical or physiological alternative to swallowing foreign matter.

The function of this foreign material is not altogether understood. The Indians of the Orinoco believe that the stones aid the crocodile in diving and in holding its prey under water, and Cott believes that they may be right. He found that the stomach stones in the Nile crocodile rather consistently make up between 1 and 2 percent of the animal's weight on land and increase its specific gravity from about 1.08 to 1.09 or slightly more. The biological value of this ballast is to permit the animal to lie submerged without being dislodged by the current and to float on an even keel, for the crocodile tends to be normally somewhat tail-heavy and top-heavy.[41] A more popular idea among zoologists is that the hard material helps to grind food which, as with nearly all reptiles, is swallowed with very little chewing. This grinding activity has been observed by x-ray cinematography.[42] Other suggestions are that the material helps to appease hunger pangs during periods of enforced starvation or supplies trace minerals not in the normal diet.

Dens and Burrows

Throughout its swampland habitat, the American alligator digs large holes, sometimes 4 to 6 feet deep and 10 to 15 feet wide. These connect with an underground passage that may be 30 to 60 feet long. Apparently only the adult animals make these refuges. They loosen the mud and vegetation with their feet and move it away by moving backward while undulating the tail. They may use their jaws to break or tear away roots but rarely for the actual excavation process. A hole and burrow large enough to hide the alligator may be completed in a few days, but the animal may continue work intermittently on its den for years. The primary purpose of the den seems to be as a winter retreat, but it is also a refuge in time of drought. Some alligators evidently never go far from their dens; others may wander several miles from them during summer and return in the fall. McIlhenny says an alligator may occupy the same den for more than 30 years.[43] Only one adult lives in a den, but newly hatched young may spend their first winter in the den of their mother. These "gator holes" are important in the ecology of the Everglades and other areas of southern marsh. Since they hold water even in the dryest periods, they serve as refuges for fish and other forms of aquatic life. The vegetation around the holes is usually more luxuriant, and some plants may also depend upon this niche for their survival.

The Chinese alligator, another crocodilian living in a region with several months of cold weather, also digs a hibernating burrow. The burrows Pope reported along the banks of the Chingshui River evidently did not contain water. He mentions that a wildcat was found in one of them, so they are at times used by other animals.[44] In the Zapata Swamps of southwestern Cuba, the American crocodile makes holes and burrows much like those made by alligators in the Everglades, although their purpose here seems to be protection from drought. The Cuban crocodile, which lives in the same swamps, is reported not to have this habit.[45]

The Indian crocodile is another species that excavates a tunnel to escape chill and drying. In February 1960, two villagers from Bahawalpur District in West Pakistan brought several baby crocodiles to the Karachi Zoo. They found them in a riverbank burrow with an adult, presumably female crocodile which they killed with a spear and hand ax. According to their account, the adult put up a fierce battle, but they may have elaborated on this in hope of getting a better price for the young. In this part of Pakistan during February, the days are pleasantly warm but nights can be miserably cold. The larger reptiles hibernate but may emerge to bask during the warmest hours. Nile crocodiles in Zululand are said to make burrows during cool weather and spend most of their time in them.[46]

Natural Enemies and Friends

Crocodiles and alligators begin their lives as small reptiles with a host of enemies; adults of the larger crocodilians are the master predators of their environment and have little to fear from any aggressor except man. Nearly all of the creatures that feed on crocodilian eggs are equally fond of a meal of hatchlings. This is true of monitor lizards, the various species of mongoose, otters, raccoons, fishing cats, and other carnivores that hunt along the waterside. Many of the larger wading birds, such as the marabou stork in Africa and the adjutant stork of India and Burma, consume many young crocodiles. Audubon painted the sandhill crane in the act of feeding on a young alligator; it is one of several

American wading birds that prey on these saurians. The big African softshelled turtle known as Thirse in Egypt is counted as one of the important enemies of young Nile crocodiles. The Florida softshell has been known to eat small alligators, and several of the big softshells inhabiting the rivers of Asia are certainly large enough to be predators on young crocodiles.

The anaconda is known to feed upon both young and near-adult caimans. I first became aware of this several years ago when a young herpetological colleague in Indianapolis, who had been quite successful in keeping a boa constrictor and several species of local snakes, decided to enlarge his horizons and spent several months' allowance on a 5-foot anaconda. For months the snake stubbornly fasted, although he offered it all the things he and I could think of that might tempt the appetite of such a reptile. One day, being short of cage space, he put the anaconda with a caiman he had lovingly raised from a pet-shop baby to a youngster of almost 3 feet. That night he phoned me and said in a melancholy voice, "Now I know what anacondas eat." Giant frogs of the Everglades have been reported to swallow newly hatched alligators, as have cottonmouths and other snakes.[47] Few enemies of young and half-grown crocodilians take a greater toll than the adults of their own species. Once the brief and uncertain period of maternal protection is over, they are fair game for any adult. For several species this leads to the immature reptiles' occupying a habitat different from that of the adults.

Once the crocodilian has reached breeding size—a matter of 6 to 10 years, depending on the species and local climatic and environmental conditions—it has far fewer enemies, although if it is a male, it may be attacked and perhaps killed by larger males. Not many animals will voluntarily tackle a crocodilian of 10 feet or more. Despite the impression gained from some popular movies and adventure stories, large wild animals avoid combat with one another. Carnivores try to get their meals with a minimum of effort and risk, while the large herbivores such as the elephant and rhinocerus are generally belligerent only in defending territory or protecting their young. The Nile crocodile and the hippopotamus share a common habitat in the lakes and rivers of Africa, where their relationship is one of armed neutrality. The crocodile is no match for an adult hippo and will generally back down if there is a confrontation. Big crocodiles are fond of eating hippopotamus calves but must avoid the protecting mother, who can cut a good-sized crocodile in half with one snap of her jaws. There are several instances of lions' killing adult crocodiles; one took place on an African game reserve in full view of a launch loaded with sightseers. In another, a crocodile attempted to steal a lion's kill and was itself slain by the big cats.[48] Jaguars are said to prey regularly on caimans, and leopards and tigers have been reported to kill and eat adult crocodiles. The ichneumon, or Egyptian mongoose, like other mongooses an important enemy of young crocodiles and a raider of nests, was said by Pliny the Elder to kill the adults as they basked with their mouths open:

> Now that he is lulled as it were fast asleep with this pleasure . . . the rat of India or ichneumon spieth his vantage and seeing him lie thus broad gaping, whippeth into his mouth and shooteth himself down his throat as quick as an arrow, and then gnaweth his bowels, eateth a hole through his side, and so killeth him.[49]

Pliny also said that dolphins attacked crocodiles and killed them by slicing the thin belly skin with their fins. This may have some factual basis in that combats between crocodiles and sharks do sometimes occur. The battle here may go either way, for there is one report of a shark found in a crocodile's stomach.[50]

One of the oldest bits of crocodile natural history is the crocodile's relation with a small bird (Trochilos). In Pliny's version:

When he hath filled his belly with fishes, he lieth to sleep upon the sands in the shore, and for that he is a great and greedy devourer, somewhat of the meat sticketh evermore between his teeth. In regard whereof cometh the wren, and she, for her victuals sake, hoppeth first about his mouth, falls to pecking it with her little bill and so foreward to the teeth which she cleaneth to make him gap. Then getteth she within his mouth, which he openeth wider by reason that he taketh so great delight in this her scraping and scouring of his teeth and jaws.[51]

Although this sounds like the sort of tale a Roman visitor to Egypt might make up to amuse the boys back at the Forum, it happens to be almost entirely true. It is, indeed, only another version of the sort of "cleaning symbiosis" that is seen on coral reefs between certain small fishes and larger ones and between fish and some kinds of shrimp. It also has a terrestrial counterpart in the relation of such birds as cattle egrets and tickbirds to large grazing animals. The recent observations of Cott and others have amply confirmed this relationship for the Nile crocodile, and there are also reports of the saltwater crocodile's being similarly attended. The bird that most frequently attends the Nile crocodile is the spur-winged plover or zic-zac, although the common African sandpiper and the water dikkop are other regular associates. The birds systematically work over the bodies of basking crocodiles as soon as the saurians emerge from the water. They feed mostly on leeches and the blood-sucking tsetse flies.[52] The common tsetse fly of crocodiles, incidentally, is one of the species that transmits sleeping sickness to man; however, the crocodile plays no part in the epidemiology of human sleeping sickness. The crocodile birds take their prey from all parts of the reptile's body and do not seem to have any special predilection for the mouth; nevertheless, observers have occasionally seen them enter crocodiles' mouths and capture food there without causing any reaction on the reptile's part.

Many other kinds of shore and water birds, such as herons, geese, ducks, ibis, and egrets, frequent the basking sites of crocodiles. The birds walk among and even over the reptiles and are never disturbed. In the water the same birds may be quickly captured by crocodiles. The birds provide the basking crocodiles with an effective early-warning system. The craning of a goose's neck, the shrill call of the spur-winged plover, or the flurry of a sandpiper will alert the reptiles and often send them stampeding into the water. There is yet another aspect to the bird-crocodile symbiosis. The water dikkop nests at the same season as the crocodile and often within 12 to 15 feet of the brooding saurian. The crocodile's presence probably protects the bird's nest as well as her own.[53]

4. Crocodilians and Man

Attacks on Man

Although Walt Kelly's comic strip and the Florida tourist bureau have given the alligator a somewhat benign image, the crocodile is still associated with remorseless ferocity and man-eating habits. In point of fact, only two species of crocodiles make unprovoked attacks on man with anything approaching regularity. However, both these species, the Nile crocodile and the saltwater crocodile, are wide-ranging and until the last two decades were abundant as well. Everywhere that these two species occur, they are recognized as dangerous, and in areas such as Thailand, West Africa, northern Australia, and New Guinea, where they are found with another less dangerous species of crocodile, the local people make a distinction between the two. Rare but authentic reports of man-eating and unprovoked attacks on man are on record for the Indian crocodile and the American crocodile. For some other large species, such as the Orinoco crocodile and Cuban crocodile, there is a lack of reliable information. Both are large enough to be put in the category of potential man-eaters, and the Cuban crocodile has the reputation of being active and dangerous in captivity. In the alligator group, the American alligator and the black caiman sometimes attack man. If historical accounts can be believed, the Chinese alligator may also have done this centuries ago when it was more abundant and grew to a larger size. Despite its great size, the gharial is one of the least dangerous of the crocodilians. There is almost no information on the false gharial, but it too is believed to be innocuous.

Unprovoked attacks by crocodilians on man are almost invariably interpreted as evidence for man-eating habits, but it is becoming clear that some of these attacks are actually defense of territory by large males. P. B. M. Jackson of the Joint Fisheries Research Organization of Northern Rhodesia reported that his boat was attacked repeatedly by crocodiles in the Mweru Marsh Game Reserve where the reptiles are protected from hunters and have little contact with man. In describing one attack he says, "The writer . . . looked up to see a large crocodile coming straight for the boat. It was swimming at a very fast pace on the surface of the water, in a flurry of spray, its tail going from side to side and its forelegs working, with its jaws opening and closing."[1] This is totally different from the silent, torpedo-like attack that crocodiles make on their prey. In this same marsh, two American biologists had a harrowing experience. As they were crossing an area of open water in a rubber boat powered by a small outboard motor, a crocodile that they estimated as over 4.5 meters (about 13 feet) long attacked from beneath the water. Its first rush burst the bow compartment of the boat, and it raked one of the occupants across the buttocks with its teeth. In the ensuing struggle, the other man was seized by the foot but managed to kick free. Luckily at this point the crocodile became more involved with the still partly inflated boat than with its occupants, and they swam to shore unmolested. When last seen, the crocodile was still trying to submerge the boat.[2] It is hard to say if this attack was motivated by hunger or territorial defense; its unexpectedness suggests the former.

The frequency of crocodile attacks varies in different parts of Africa. Major Flower, writing in 1933, said that crocodiles were fairly numerous along the main stream of the Nile in Dongola

Province, Sudan, but he knew of only one fatality in a seven-year period. On the other hand, in the Fourth Cataract country, only a short distance to the south, fourteen people in one village alone were killed or badly mauled by crocodiles during 5 years.[3] George Cansdale, a British zoologist with 14 years' experience in the Gold Coast of West Africa, wrote that he could personally authenticate only one crocodile attack that resulted in serious injury. He added, "There must be other cases but I found them hard to check."[4] In his ecological study of the crocodile in Uganda and Rhodesia, Cott recorded stomach contents of 851 specimens taken between 1952 and 1957. Of these, 444 were 2 meters (78.5 inches) or more long and thus in the size range for potential man-eaters. Human remains were found in only four stomachs, or less than 1 percent of those examined. Cott also mentioned that a game ranger was killed by a crocodile but not eaten.[5] The finding of human remains is not, of course, absolute evidence of predation on man. All crocodilians, so far as is known, are to some degree scavengers, and the dead body of a man would be eaten as readily as that of any other animal of comparable size. In India and some other parts of southeastern Asia where religious ritual consigns bodies of the dead to rivers after partial cremation, gharials, Indian crocodiles, and saltwater crocodiles have been said to utilize this food source. "When will crocodiles refuse corpses?" is a Malay saying implying that some people will take anything that comes their way by chance. But even if the 1-percent figure for the Nile crocodile in central Africa represents true predation on man, it probably does not justify the statement that crocodiles cause more human casualties in Africa than all other wild animals together (excluding poisonous snakes).[6]

Although the saltwater crocodile is also an undoubted man-eater and perhaps a more dangerous one than the Nile crocodile, there is less reliable information regarding the loss of life it causes. Unprovoked attacks on man resulting in death or serious injury have been reported from all parts of the species' extensive range. Most often they seem to be isolated instances described for their dramatic and admonitory value. However, the bloodiest crocodile attack reported in recent times took place during the last months of World War II. A detachment of Japanese troops trapped on Ramree Island just off the west coast of Burma attempted to escape across a tidal swamp separating the island from the mainland. The swamp was heavily populated by saltwater crocodiles that moved in to attack after dark. Nearly a thousand Japanese soldiers were said to have perished that night, but there is no breakdown as to how many died from gunfire, drowning, and other causes and how many were killed by crocodiles.[7] This is the only reasonably well authenticated account of a mass attack on man by crocodiles. It would be interesting to know the previous history of this crocodile population's relationship with man. One would surmise that they had been man-eaters that had not been seriously disturbed by hunting and were uncommonly bold. A crocodile attack with a happier ending than most involved Dr. Philip J. Darlington, an American zoologist well known for his outstanding work in zoogeography and entomology. During World War II, Dr. Darlington was engaged in malaria-control work on the island of New Britain. As he was collecting mosquito larvae in a swamp, a crocodile about 10 feet long surfaced near him. As he tried to retreat, he slipped off the log on which he was standing, and the crocodile seized him by the arms. In the manner of its kind, it dragged him under water with a violent twisting motion. Darlington was a powerful man in good physical condition. After a seemingly endless struggle, the reptile released him, he believes after he landed a solid blow on its belly. He suffered a dislocation and compound fracture of his forearm and several tooth punctures.[8]

During the time he was in New Guinea, Wilfred Neill saw one man carried away and presumably eaten by a saltwater crocodile and another man's foot torn off by a crocodile that attacked him while he was in a canoe. Neill heard of numerous other injuries and fatalities in that region.[9] When the young anthropologist Michael Rockefeller disappeared off the southern coast of New Guinea in 1961, several people with knowledge of the area felt that he might have fallen victim to a crocodile. From northern Australia come other reports of fatalities and injuries, but most of them seem to have been sustained during crocodile hunts when injured or cornered animals retaliated in self-defense.

The Indian crocodile has a generally good record with respect to attacks on man, although individuals of the subspecies inhabiting Ceylon are said to become ferocious man-eaters occasionally.[10] As is true of tigers and other big cats, man-eating by crocodiles is often believed to be an aberration of certain particular animals. This may well be true, especially where man is one of the more readily available prey species, but the life of a man-eater is apt to be a short one today with power boats, electric jacklights, and high-powered rifles widely available. It is sometimes said that in India crocodiles acquire a taste for human flesh by eating corpses placed in the rivers after partial cremation, but whether the crocodilian palate is that discriminating is open to doubt.

Early accounts of the crocodile shrine near Karachi mention pilgrims and visitors killed by the religiously protected reptiles. Even if true, these represent reactions of animals living in an abnormal relationship with man. During the more than four years we were in Karachi, we heard no account of human death or injury due to a crocodile in any part of West Pakistan. The newspapers carried an occasional report of such attacks from the coastal regions of East Pakistan, where the saltwater crocodile was doubtless the species at fault. The lake and river people of Sind, who were well acquainted with the Indian crocodile, or *wugu*, and the gharial, or *say-sar*, had no particular fear of either species, but this was not true of the farm and village folk who knew the reptiles largely by hearsay.

The American crocodile can unquestionably be dangerous under some circumstances. In 1925 a party of surveyors shot a large crocodile on the shore of Biscayne Bay near Miami. One man walked up to the seemingly dead animal and kicked it, whereupon it suddenly revived, dealt him a terrific blow with its tail, and mauled him so severely that he died. This reptile was captured by men from a local alligator farm and later acquired by Ross Allen's Reptile Institute. Here it lived until 1953, when it was killed by a huge alligator in an adjoining pen. This crocodile was estimated to be 65 years old at the time of its death. It is one of the very few man-killing reptiles to have survived to old age as a tourist attraction. The Mexican herpetologist Miguel Alvarez del Toro reports that in lowland Chiapas women and children are frequently attacked by crocodiles when they go to the rivers to wash or get water. Most attacks take place at dusk or after dark. Sometimes certain pools or fords cannot be used until the offending saurian is killed or driven away.[11] There is a report of a crocodile's killing a 10-year-old boy in Panama,[12] but such incidents seem to be few. Crocodile attacks, even if they result in nothing worse than a severe fright, are good material for village gossip and are apt to be repeated with embellishments until one gets the impression that the reptiles are a major menace.

Early travelers in the southeastern United States regarded the American alligator as a dangerous animal. William Bartram wrote: "Behold him rushing forth from the flags and reeds. His enormous body swells. His plaited tail brandished high floats upon the lake. The waters like a cataract descend from his opening jaws. Clouds of smoke issue from his dilated nostrils. The earth trembles with his thunder." Bartram goes on to say that his boat was attacked by three alligators, and he narrowly escaped with his life.[13] John Edwards Holbrook, a South Carolina physician whose *North American Herpetology* (1842) is comparable to Audubon's great work on the birds of the continent, wrote: "Alligators will, however, defend themselves boldly when on land and at certain seasons of the year . . . yet on these occasions I have never known them the aggressor."[14] In the course of many years' field experience with alligators, McIlhenny said that he was only once the target of an unprovoked attack. In this case, a large alligator slid off the bank as he approached in his pirogue and waited in the water until it could strike the boat with its tail.[15] Ross Allen, whose experience has probably been even more extensive, says that he was never subjected to a completely unprovoked attack. However, between 1948 and 1971 there were at least seven such attacks by alligators in Florida. In five of these, the victim was swimming, while two involved children who were dipping for minnows with a net when attacked. None were fatal, though all resulted in moderately severe injuries. The alligators were reported as being 6 to 12 feet long, yet all either released their victims when they struggled or were driven off with little difficulty.[16] It almost looks as if all these attacks were the result of the reptiles' mistaking a human for their more normal prey. If one discounts some of the accounts of early explorers, man-eating by alligators seems to be extremely rare. Holbrook said, "There is, I believe, no well authenticated instance with us in Carolina, of their having preyed on man; yet Lacoudrenière says it often happens in Louisiana, and that they greatly prefer the flesh of the black to the white!"[17] There has been one well-documented recent incident, however. In 1957 a 9-year-old boy disappeared from his home at Eau Gallie, Florida. Three days later his mutilated body was found in nearby Horse Creek, and the missing parts were recovered from the stomachs of a pair of adult alligators killed nearby. It is not certain whether the alligators killed the boy or found his body and fed on it; the medical evidence was equivocal.[18]

Part of the reason for alligator attacks in Florida is the encroachment of a rapidly growing human population on the alligator's habitat, plus the protection of the big reptiles, at least in some places. Conservation of the alligator and the world's other crocodilians is unquestionably a worthwhile thing, but it must be realized that most of these animals are potentially dangerous and may readily lose their fear of man. On the popular Anhinga Trail in Everglades National Park we recently saw four large alligators and one small one within the space of no more than a half-hour. The reptiles were practically indifferent to us and to other tourists gawking at them from a distance of a few yards. The behavior of such animals toward man is apt to be more unpredictable than that of animals whose contact with man is infrequent; moreover, the manifold imbecilities of the human race can lead to trouble. A few National Park visitors seem to think the native wild animals are large mechanical toys activated for their entertainment or, as we heard in Yellowstone, "rangers dressed up like bears." When the animals act like animals, someone is likely to be hurt.

Early explorers in South America referred to man-eating by both crocodiles and alligators, but it is impossible from their accounts to know what species they had in mind; the American crocodile, the Orinoco crocodile, and the black caiman seem to be likely candidates. Don Antonio de Ulloa in 1735 wrote:

> These alligators are great destroyers of the fish in this river. . . . When they cannot find fish to appease their hunger, they betake themselves to the meadows bordering on the banks and devour calves and colts. . . . There are even too many melancholy instances of their devouring the human species, especially children, who from the inattention natural to their age, have been without doors after it is dark. . . . Their voracity has also been felt by the boatmen, whom, by inconsiderately sleeping with one of their arms or legs hanging over the side of the boat, these animals have seized, and drawn the whole body into the water. Alligators who have once feasted on human flesh are known to be the most dangerous, and

> become, as it were, inflammed with an insatiable desire of repeating the same delicious repast.[19]

It is likely that these accounts were obtained second-hand and accepted uncritically. Zoologists with more recent experience in South America have not reported casualties from crocodilians except in rare instances, most of them among the professional crocodile hunters, or *caimaneros*.

Crocodile Hunting

Wherever crocodilians are found, men have hunted them from earliest times. Until the development of modern firearms, the contest was not too unequal. Many of the older methods are much like those used for catching large fish. In Malaya, a piece of wood sharpened at both ends is attached to a raveled length of rope, the strands of which pass between the crocodile's teeth and prevent the rope from being bitten in half. This is fastened to a rattan line 20 to 60 feet long that is in turn usually secured to a pole. The sharpened stick is concealed in the body of a duck or chicken that is then placed on a small raft. If a crocodile takes the bait, the stick becomes lodged in his gullet. He will usually swim to shelter but is located by the floating pole and rattan. The hunter in a boat plays the crocodile at the end of the line until it is exhausted; then it is killed by cutlass strokes or with a gun. The entire process of making and baiting the trap, waiting for the crocodile to take the bait, and finally pursuing and dispatching the animal is accompanied with elaborate ritual and magic, particularly if the crocodile is a man-eater. The sharpened stick must come from the bottom rung of a house ladder and the pole must be cut by three strokes of the right hand. The bait is bound with rattan seven times round and seven times across and a mixture of betel leaf, ginger, and peppercorns is chewed up and spat onto it. As the baited trap is placed in the water, a long charm is recited:

Ho, Si Jambu Rakai, I know your origin;
Sugar-cane knots forty-four were your bones,
Of clay was formed your body;
Rootlets of the areca palm were your arteries,
Liquid sugar made your blood,
A rotten mat made your skin,
And a midrib of the thatch palm your tail,
Prickles of the pandanus made your dorsal ridge,
And pointed *berembang* suckers your teeth.
If you splash with your tail it shall break in two,
If you strike downwards with your snout it shall break in two,
If you crunch with your teeth they shall all be broken.
Lo, Si Jambu Rakai, I bind this fowl with the sevenfold binding,
And enwrap it with the sevenfold wrapping
Which you shall never loosen or undo.
Turn it over in your mouth before you swallow it.
O Si Jambu Rakai, accept this present from Her Highness Princess Rundok from Java.
If you refuse to accept it,
Within two days or three
You shall be choked to death with blood,
Choked to death by Her Highness Princess Rundok from Java.
But if you accept it,
A reach upstream or a reach downstream, there do you await me;
It is not my word, it is King Solomon's Word;

If you are carried downstream, see that you incline upstream,
If you are carried upstream, see that you incline downstream,
By virtue of the Saying of King Solomon, "There is no god but Allah."

While the hunter is waiting for the crocodile to take the bait, he must begin his meal by swallowing three lumps of rice and must not remove the bones from the meat in his curry; otherwise the crosspiece will work loose and the crocodile will escape. After the crocodile is hooked, the hunter recites the "crippling charm," which again recounts the crocodile's origin and adds:

Peace be with you, O Prophet Khailir,
In whose charge is the water;
Peace be with you, O Prophet Tetap,
In whose charge is the earth;
Pardon, King of the Sea, God of Currents,
I wish to drive the mischief out of this crocodile.[20]

In Borneo, the hunting of a man-eating crocodile is entrusted to a magician summoned by the bereaved family. After a price has been agreed upon, the sorcerer prepares himself for the hunt. He becomes ritually unclean and may not pass in front of doors or windows. He cooks nothing for himself but depends on charity. He must swallow his food unchewed and must not prepare any fish or nuts or peel any fruit. He builds a tiny hut at the water's edge near where the victim was caught and there prepares a special boat painted red and yellow and fitted with lances. Then he casts lots to determine what kind of flesh to use for bait. Finally he baits his hook and pushes off in his boat chanting: "Ye crocodiles who are upstream, come down, and ye crocodiles who are downstream, come up; for I will give you all good food, as sweet as sugar and as fat as coconut. I will give you a pretty and beautiful necklace."

When a crocodile takes the bait, the magician calls, "Choose a place for yourself where you will lie; for many come to see you. They come joyfully and exultingly and they give you a knife, a lance, and a shroud!"

If the crocodile's stomach shows no trace of human remains, the hunter continues to catch more crocodiles until he finds one that does. When he has shown the victim's family proof of catching a man-eater, the men then sacrifice a cat to the remaining crocodiles as a ritual apology for having disturbed them.[21]

The elaborate ceremony involved in Malaya and Borneo indicates that hunting these big and dangerous reptiles was fraught with peril and uncertainty, and the hunter obviously tried to make it clear that his enmity was toward a particular animal and not to the whole saurian tribe.

Variations of the baited-hook technique are used for catching crocodilians in other parts of the world, including South America and New Guinea. We have been told that it is currently popular among United States alligator poachers because it is less obtrusive than hunting with a rifle and jacklight, and it is virtually impossible to prove that the lines were not set for catfish, gar, or some other legitimate quarry.

An alligator trap popular with the Seminole Indians in former times was made of two converging rows of stout stakes hammered into the mud. At the apex was a narrow, S-shaped passage lined by rows of similar closely placed stakes. Bait was placed at the end of this. The alligator entered the trap and became wedged as it tried to back out. If it was not needed immediately, it would remain alive in the trap for many days.[22]

The method used by the river people of Sind to catch the Indian crocodile illustrates the basically timid nature of this beast. When a crocodile has been sighted, a group of twenty or so men with large-meshed nets and bamboo poles attempt to surround it. Each man supports himself on a float, or *dillo*, which is a round copper vessel with a hole in the top. The hunter lies on the float, plugging the hole with his belly, with his limbs free to paddle and to use a pole or net. The crocodile generally responds to the disturbance by going to the bottom and working its way into the mud where it is located by probing. If it moves, the hunters try to follow the trail of bubbles it leaves and again surround it with their nets. When it finally comes to rest, they fix four bamboo poles on each side of its neck and body to hamper its movements. Then two of the more experienced hunters dive down and attach a rope snugly around the crocodile's body. As the rope is pulled, the crocodile tries to fight its way free but becomes entangled in the nets and is hauled ashore. It is then killed with the little long-handled axes that most Sindhi men carry, or with a gun if one is available.[23]

Crocodilians frequently become entangled in fishnets. The largest alligator taken alive in the last 50 years was caught in a fishnet in Lake George, Florida. Until about 1950, many American crocodiles were netted by fishermen working out of the Florida Keys. More often than not, large crocodilians escape from the nets, thoroughly wrecking them in the process. In many parts of Asia and Africa where fishnets are laboriously made by hand, their destruction means many hours of extra work and a curtailment of the village food supply. On the other hand, natives of the estuary of the Volta in West Africa rely on crocodiles to drive fish into their traps, which consist of rings of stakes surrounding a pile of branches. Fleeing the crocodiles, the fish hide under the branches. When enough fish have gathered, the crocodiles are driven away, and the fishermen enclose the stockade with a split-cane fence, toss out the branches, and easily catch the fish in scoop nets.[24]

Pliny mentioned that a tribe of short-statured men who lived on an island in the Nile killed crocodiles by thrusting a sharpened stick vertically into the opened mouth.[25] This has also been reported from other parts of the world. Considering the activity of the reptiles and the force with which they close their jaws, it sounds like a highly hazardous practice.

Spearing or harpooning crocodiles is a widespread method of hunting in Latin America, Africa, Australia, and New Guinea. It may be done either by day or night and takes considerable skill, muscle, and nerve. The horny back of a big reptile will turn all but a well-placed harpoon, and its jaws and tail make it a formidable adversary when it has been played out and brought alongside the boat. This method is preferred over shooting by scientific collectors, because the skull, which is important in taxonomic studies, need not be damaged, while it is often shattered if the animal is shot. On the other hand, spear holes decrease the market value of the hide for the leather trade.

Shooting crocodilians is apparently almost as old as the invention of firearms. Until the nineteenth century, guns were inaccurate and clumsy to load, and their muzzle velocity was so low that they were no more effective against the bigger reptiles than more primitive methods. With the development of more accurate and faster-firing guns and more powerful ammunition, the crocodilian armor was no longer a defense. Shooting alligators from river steamers had made these reptiles rare along the major rivers of the southern United States by 1890. The same thing happened, although more gradually, in India, southeast Asia, Africa, and South America. In the twentieth century the development of small powerboats and strong electric spotlights gave the hunters ready access to spots where the reptiles had previously been relatively secure and facilitated night hunting. Like many amphibians and reptiles, crocodilians can be approached more closely at night. When Philip W. Smith, a herpetologist and ichthyologist with the Illinois Natural History Survey, and I were collecting in northern Florida in 1954, we saw no sign of alligators by day. However, in the same places after dark we caught the red reflections of several pairs of eyes, some disconcertingly large, and easily picked up a small alligator lying in shallow water. Shooting is most effective if the only motive of the hunter is to kill. If the animal is wanted for its hide or for some other purpose, it may be lost, especially if it is shot in the water. Nevertheless, the usual method of professional hunters in Africa is to work at night in a fast motorboat, locate the crocodile by eye-shine, shoot it at point-blank range, and gaff the dying animal before it can sink. Using these tactics, hunters have virtually exterminated crocodiles in many places where they were numerous prior to 1950.

Utilization by Man

Commercial hunting of crocodilians is chiefly for their hides, which yield an attractive and durable leather. All species can be utilized. However, the extensive development of the osteoderms in some of the caimans make their skins difficult to process, and some other species are too small or too rare to be of commercial value. Only the skin of the belly and flanks can be made into leather; the dorsal skin is too heavily armored to be used. To give some idea of the numbers taken, in Uganda in 1953 the crop was 15,000 skins valued at £100,000; in 1954, 7,900 at £44,553; in 1955, 8,000 at £40,000. In the whole of East Africa for that period, about 60,000 skins were taken annually. These data are based on export permits granted and do not represent the total take.[26] In Florida in 1929 top-quality alligator hides brought $1.50 each, and 190,000 passed through dealers' hands. In 1934 the price had doubled and the number of hides dropped to 120,000. By 1939 the price stood at $5.25, but only 80,000 hides were secured. After this, there was a rapid decline in the numbers of alligators, and in 1943 only 6,800 hides were taken, despite a price of $19.25 for 7-foot hides and more for larger ones. In 1944 the species was protected during its breeding season and the taking of individuals under 4 feet was prohibited. The effect of this legislation was rapidly apparent; in 1947, 25,000 hides were brought in, and the price dropped to $13.30.[27] Alligators fared reasonably well in Florida during the 1950s, but subsequently an increased demand for reptile leathers has led to widespread poaching. This has been combined with a marked reduction of suitable alligator habitats coincidental with the rapid growth of Florida's human population.

Hides of the saltwater crocodile and New Guinea crocodile are the basis of an industry that earned nearly a million dollars and contributed about 2 percent of the export income of the Territory of Papua and New Guinea between 1963 and 1966. In 1969 this had fallen to less than a half-million dollars and 0.6 percent, in spite of an increase in the value of top-quality hides from $2.50 per inch of belly width in 1965 to $3.30 in 1969. A 1970 report indicated that the industry would continue to decline unless controls were enforced to protect the breeding stock in the wild or some type of crocodile farming instituted.[28]

In the central Amazon region, early European explorers reported that caimans were as plentiful in the rivers as tadpoles in an English ditch, and as recently as 1930 a caiman was not considered worth the ammunition to kill it. Since 1940 the demand for hides and for young caimans for the pet trade has led to the virtual extinction of the black caiman and a great reduction in numbers of the spectacled caiman. The surviving reptiles have become extremely shy, and the sight of one is unusual.[29] In the Paraná drainage of southern South America, the Paraguay caiman and the broad-snouted caiman have suffered a similar fate.

In northern Australia saltwater crocodiles, formerly abundant in estuaries, were so reduced in numbers between 1945 and 1960 that hunting became no longer financially profitable.[30] The American crocodile has also suffered severely as a result of the leather trade and is rare or extinct in many places where it was formerly plentiful.

Crocodilians provide a substantial amount of palatable meat and have been hunted for this reason in many places. They are a regular part of the diet of the blackfellows in northern Australia and New Guinea. In southeastern Asia they are eaten by the Burmese, Chinese, and some other ethnic groups. Crocodiles are rarely eaten in the Indian subcontinent. In Africa, their flesh is prized by some tribes and completely avoided by others. Alligator was a regular part of the Seminole Indian diet until quite recent times and is still probably eaten occasionally. In Holbrook's time, alligator tail was part of the slaves' ration in Carolina. Holbrook himself sampled it and pronounced it "tolerable."[31] Until about 1950 alligator soup was produced by a Florida cannery and distributed in gourmet food shops throughout the United States. The Indians of Guiana organize caiman hunts two or three times a year when the water level is low in the swamps. The reptiles are rounded up like cattle and killed with axes and clubs. Meat of the spectacled caiman is relished; that of the black caiman is said to be unfit to eat.[32]

Alligator hunters sometimes take the teeth of their quarry, which are subsequently sold as charms or souvenirs. In an early (1882) plea for the conservation of the alligator, it was said that one man in Florida had collected 350 pounds of alligator teeth.[33] In Rome of the first century A.D., the eye tooth of a crocodile filled with frankincense and tied to any part of the body was a cure for periodical fevers, provided that for the next five days the patient did not see the person who tied the amulet to him. A tooth from the right side of a crocodile's jaw was bound to the right arm of either sex "to provoke unto carnal lust."[34]

Prince Maximilian of Wied-Neuwied, who traveled extensively in both North and South America during the 1820s and

1830s, reported that ladies of Spanish colonial America made a perfume for their hair from rosewater and alligator musk.[35] Similar use of crocodilian musk is known from other parts of the world.

The gall bladder of the saltwater crocodile is highly prized in Thailand as a cure for sterility in women;[36] and Marco Polo mentioned that the gall of what was presumably the Chinese alligator was a valuable medicine. In Africa, however, this substance is considered a most virulent poison. In some tribes, the gall bladder of any crocodile killed becomes the property of the chief, because it is too potent a substance to be permitted to fall into the hands of an ordinary individual.[37]

While millions of crocodilians have been killed for their hides, meat, and other products of real or presumed value, many more have been killed for sport or to protect livestock and human beings from their attacks. In the great caiman hunts at the mouth of the Amazon between 1900 and 1908, up to 300 or 400 animals a day were killed in what was primarily a campaign of extermination. An account of life in the Panama Canal Zone soon after the construction of the canal says that killing alligators (probably American crocodiles) was a popular sport.

> If you visit Crocodile Creek with a typical party you will be given a very fair imitation of a lively skirmish in actual war. From every part of the deck, from the roof of the cabin, and from the pilot house shots ring out from repeating rifles in a fierce desire to kill. The Emersonian doctrine of compensation is often given illustration by the killing of one of the hunters in the eagerness to get at the quarry. In fact, that is one of the commonest accidents of the tourist season in Panama. . . .
>
> Just what gives the killing of alligators its peculiar zest I could never understand. The beasts are slow, torpid and do not afford a peculiarly difficult test of marksmanship, even

> though the vulnerable part of their bodies is small. They are timid and will not fight for their lives. . . . They are practically harmless, and in the Bayano region wholly so, as there are no domestic animals upon which they can prey. It is true their teeth and skins have a certain value in the market, but it is not for these that the tourist kills them. Most of those slain for "sport" sink instantly and cannot be recovered.[38]

Even the reptiles' friends have not been without fault. Although E. A. McIlhenny established a refuge for the alligator on his lands in southwestern Louisiana and was one of the first persons to work for its conservation, he tells how he once shot more than eighty alligators in a few hours in a pond where one of the reptiles had killed one of his favorite retrievers.[39]

Alligator wrestling is a distinctively American development, the counterpart of snake charming in India. Neither performance is as dangerous as the promoters would lead one to believe, but both have a genuine element of risk. Alligator wrestling is probably the more hazardous, for most of the snakes used by snake charmers in the East have been defanged, while the alligators are in full fighting trim. Although many alligator wrestlers today are Seminoles, there is no evidence that this is any part of their tradition. In fact, the first alligator wrestling matches seem to date back no further than the Florida land boom of the mid-1920s and originated among operators of the alligator farms that were popular tourist attractions of the time.[40] Aside from presumably superior intelligence, the man has at least two other advantages over his saurian opponent. While a big alligator can close its jaws with strength enough to break the leg bones of a cow, the muscles that open the jaws are just strong enough to carry out this function. A man can easily hold shut the jaws of even a very large alligator with his hands, and a common finale for an alligator-wrestling match is for the wrestler to pin the saurian's jaws between his chin and chest. The second advantage is that the crocodilian's heart and wind are not designed for a prolonged struggle, and a well-trained athlete can nearly always outlast a reptile. There have been several instances when the human wrestler accidentally drowned his saurian opponent during a match. Finally, an alligator probably sees very little sense in fighting a man for the sheer hell of it and is usually not a particularly savage antagonist. Things can go wrong, however, and some alligator wrestlers have been severely injured. It is said that 8- to 9-foot alligators are most dangerous because they combine strength with agility; the bigger animals tend to be slow and clumsy.

Conservation

Until quite recent times, the general attitude of the peoples of Europe and America was that crocodilians were ugly, dangerous, and quite without redeeming qualities. They were generally killed whenever and wherever found, on the assumption that their destruction would be to the public benefit and the world would be a better place without them. Only in the past few decades have there been serious attempts to assess their role in the balance of nature and their economic value to man. The judgment has been far more to their favor than to their disfavor, and serious attempts for their conservation are being made, albeit a bit belatedly for some species.

Hugh Cott's conclusions from his study of the Nile crocodile are unequivocal:

> Crocodiles are not detrimental to fishery interests, except insofar as . . . they damage gear. On the other hand, there is

evidence that their presence is directly or indirectly beneficial to the industry. . . . The specious argument that the Nile crocodile merits persecution because it endangers human life will impress no one who is familiar with the species in the field today. Its depredations have often been grossly exaggerated. . . .

There are two grounds for regarding the crocodile as a commercial asset. Firstly, it is a producer of high quality leather. . . . Under rational management, the reptiles could provide a sustained yield of skins to the trade. . . . The species has a subsidiary, but by no means negligible value, as a tourist attraction. . . .

Crocodilians also merit protection in their own right. . . . As the only remaining members of the archosaurian stock . . . they are of quite exceptional scientific importance. . . . It would be a grave loss to science and research, and to posterity, if these saurians—which have survived for over a hundred million years—were now to be sacrificed to the demands of uninformed public opinion. . . .

The general problem of the Nile crocodile's relation to man is a very old one: ". . . Shall the bands of fishermen make traffic of him? Shall they part him among the merchants?" It is hoped that the authorities concerned will now reconsider the question . . . and so take appropriate steps before it is too late to save this unique and valuable member of the African fauna.[41]

At the present time the Nile crocodile is one of five crocodilian species on the U. S. Department of the Interior's list of rare and endangered animals; the others are the Cuban crocodile, Orinoco crocodile, Morlet's crocodile, and the Paraguay caiman. These species may no longer be imported into the United States except by special government permit. The American crocodile and

American alligator are considered rare and endangered native species and receive at least nominal protection in all states where they occur.

Effective conservation of crocodilians must take into account the slow growth of these reptiles. In waters where adult animals have been virtually eliminated by hunting, a rest of at least 15 years is needed in the case of the Nile crocodile to re-establish an adequate breeding stock. American alligator populations seem to have more rapid powers of recovery. A marked increase in their numbers was noted within 5 years after the establishment of a closed season in Florida in 1944. The enforcement of regulations against commercial hunting is difficult because of the large areas of difficult terrain involved and the tendency, at least until recently, to let poachers off with a warning or with fines so light they could be recouped in a successful night's hunting. Poachers have operated successfully in the Everglades National Park in Florida and in Murchison Falls National Park in Uganda as well as in other protected areas. Depriving the commercial hunter of his market by legislation against the use of crocodilian leather may be the best measure for crocodilian conservation but it must be international in scope for best results. There is also the chance that such legislation will cause hunters to turn to the large snakes and lizards as sources of reptile leather. Most of these species are just as vulnerable as the crocodilians to intensive hunting.

Crocodilian farming is a possible answer to the demand for leather and is being tried with some indication of success in Florida, Thailand, Zululand, Jamiaca, and perhaps elsewhere. Most of the early alligator farms, like snake farms and turtle farms, were simply places where wild-caught reptiles were confined until used for hides, meat, or venom production; incidentally they served as lucrative tourist attractions. Now some efforts are being made to breed crocodilians in captivity and raise the young to a size at which they can be used commercially. As yet these enterprises have not been able to compete successfully with the hide-hunters. Cannibalism and the strong tendency of male crocodilians to fight among themselves are serious problems in commercial crocodilian culture.

5. Myths about Crocodilians

Mister Peacock

In the Karachi bazaar lepers beg on Thursday. On other days they generally keep off the street, and many are confined in a leprosarium some 10 miles south and west of the city. Here at Mangho Pir, dust-colored hospital buildings rim a dense circle of palms and scrub where a warm-water aquifer rises near the surface. The hospital is pitifully understaffed; in 1958 there was one resident physician, himself a leper. But modern medical treatment is supplemented by the presence of sacred crocodiles. Just across the road to the south, the warm spring waters break out at the base of a small, untidy mosque and spill into a large walled tank containing a few tiny islands. Within the tank some two dozen big Indian crocodiles float motionless in the greenish water or lie basking in the sun.

Sherman and I went often to Mangho Pir to lean on the tank's wall and speculate on our chances of getting a newly hatched crocodile for our collection. We were told that young crocodiles hatch within the tank every year, usually during June or July, but soon disappear. The presumption is that they are snapped up by the bigger reptiles, although some probably get out through cracks in the wall and may eventually find their way to the Hab River, a mile or so away.

Once we visited the shrine at the time of the Bakra Eid, a holy festival celebrating Abraham's sacrifice. The mosque grounds and the small graveyard were filled with colorfully dressed celebrants.

Down in the crocodile tank a gaunt man danced about the flotilla of reptiles, waving the bloody thighbone of a goat and rapping the crocodiles on their heads to stir them up and give the crowd a good show.

Apparently sensing our interest, a man standing near us by the wall explained that a holy saint "had long past come walking over the sand and now it is he is with us," he concluded grandly, waving his hand toward the mosque to our right.

"Always it is he is here, peace be on him!" He made a hasty salaam and turned back to us at the wall. Below, a venerable crocodile was separating itself from the rest and hauling toward the master of ceremonies. "There he is himself! The Moor-Sahib!" our informant continued.

I was puzzled. "Why Moor-Sahib?" I wanted to know.

The man shrugged. "Because he was a peacock," he replied enigmatically.

I looked down at the huge saurian, now just below me, and thought I had never seen anything that looked less like a peacock. I also noted that the beast looked hungry and tossed a few annas to the frenetic man who had begun to dance away from the advancing beast. He scooped them up with a salaam as my new friend continued, "Once it is that he was in the mosque"—he waved toward the building—"in a box with a hole in the top, and they fed him, you know, through the hole, but he was not happy." He inclined his head sadly, watching the Moor-Sahib. "He was too much lonesome, you know, so now he is here with us, isn't it?" he concluded brightly.

I watched the man below as he tossed a bloody tidbit into the waiting jaws that snapped shut with a decisive crunch. A red mark had been painted between the protuberant eyes, something I had not seen on previous visits, and I decided that this must be in honor of Abraham's festival. But I was still puzzled. The name *moor* certainly means peacock in Hindhi, but why a peacock? The explanation may be hopelessly lost among the centuries of foreign invasions and shifting cultures that have marked the history of the Indus delta. But we wonder if Mangho Pir, like many desert springs a natural holy place, may once have been sacred to Sarasvati, the wife of Brahma, a mother goddess who rode a peacock and was skilled in healing arts. Peacocks may once have lived on the oasis with the crocodiles and probably made up part of the reptiles' food. Over the centuries the reptiles would thus have obtained the spirit of the sacred fowl, and eventually their name as well.

Mangho Pir seems to be Karachi's most durable tourist attraction as well as one of the world's oldest crocodile shrines. It first appeared in Western literature in 1838 when a Lieutenant Carless of the Indian navy reported seeing a great host of crocodiles moving freely about an extensive swamp. He mentioned no tank nor enclosure.[1]

In 1854, Dr. Henry Gould, a young British physician, recorded his visit to the shrine in a letter to his father, who entered the entire account in the *Proceedings of the Zoological Society of London:*

> I will now give you a short account of an excursion to the well-known spot called "Munghur Peer" or the Alligator Tank. . . . From most of the districts of Scinde, the natives make pilgrimages to these beasts, throwing goats to them as peace offerings; the inhabitants of the small village adjoining the mosque feed them regularly and are great in the odour of sanctity. Should a small urchin be swallowed, through his indiscreetly venturing too near, they say his is a happy fate for he is sure of paradise. The brutes do not confine themselves to the water but wander among the palm trees and low bushes and wherever a sahib discovers a cool retreat, there a large alligator is sure to be found; and before a transfer of seat can be effected, must be fought and dislodged, or run away from

> if he charges. Not long since, one of them killed a man, whereupon two officers resolved to take vengeance, and accordingly, in spite of military prohibition, shot the delinquent, whereupon all the crazy and bigoted inhabitants turned out to avenge the insult and had they not been well armed, it would have gone hard with the sacrilegious scamps. These holy reptiles are not without their royal family—one lineal descendant by the name of moor-sahib who is the acknowledged king of the alligators; although superior to his subjects not only by his descent but by his enormous size and appetite. . . .[2]

Crocodiles were once esteemed in other places throughout old India from Karachi to Dacca. Hugh Wilkinson, a British traveler, in 1880 reported seeing a great hoard of these reptiles in a lake near Amber, the royal city of the Jaipur maharajas. He said that the reptiles "were considered sacred until they became too fond of the lean humankind here. When they are chased and killed, the anklets, bangles, nose and toe rings found inside being returned to their late proprietor's sorrowing families."[3]

It is doubtful if the crocodiles and large turtles kept in shrines throughout India and Burma should be considered truly sacred animals. They are not believed to be gods or even representatives of gods but rather living witnesses of past events or associates of particularly holy personages. Of course, beneath this attitude may be undercurrents of older, animistic religion.

Fatima's Favorite Toy

According to an Islamic legend from Malaysia, the first crocodile was created as a toy for Fatima, Mohammed's daughter. Her nurse, Putri Padang, fashioned the reptile from clay, making its eyes from bits of saffron and its tail of a betel frond. Then she prayed that God might give the figure life and it at once began to live and breathe. It became Fatima's favorite toy and she played with it for many years. As Putri grew old, she complained to her charge that the creature was growing unreliable and crafty, and finally Fatima cursed it, saying, "Thou shalt be the crocodile of the sea; no enjoyment shall be thine, and thou shalt not know lust nor desire."

Fatima then pulled out the crocodile's teeth and tore out its tongue and drove nails through its jaws to fasten them. But the animal managed to open its jaws and the nails have served it as teeth to this day.

As a consequence of this early mutilation, Malay Pawangs who wish to snare a crocodile bait their line with a white fowl and as they set it utter the incantation:

Kun kata Allah sagaya kun kata Mohammed tab Paku [Blood saith God, so Blood saith Mohammed: nail be fixed]!

If the incantation should fail and the hunter be seized by the crocodile and drowned, the reptile, after a suitable interval in memory of its first mistress, will return the corpse to the surface and call upon the sun, moon, and stars to witness that it is not guilty of the homocide, saying: "It was not I who killed you, it was the water which killed you." After which the crocodile may feel free to make a meal of the body.[4]

Ancient Egypt and Contemporary Africa

The crocodile as a god first appeared in Egypt, where as Sebek, personifying evil, he was a relict of prehistoric animistic cults of the Nile delta. "What won't these mad Egyptians use for

Gods?" Juvenal sneers in his Fifteenth Satire. "One district worships the green crocodile, another ibises gorged with serpents. . . ."[5]

In Egypt's baffling and complex system of animal worship the crocodile was both revered and detested: served and nurtured by priestly cults in some communities; ruthlessly slain and eaten in others. In carvings on tombs or mastobas dating back to the Fourth Dynasty (2613–2494 B.C.) Sebek in his saurian form swims under the funerary barges of pharaohs bound for the underworld. A prayer in the Book of the Dead pleads, "Oh, deliver me from the crocodiles of this land!" The hieroglyphic sign for black and darkness was a crocodile's tail.[6]

As daemon of the night, Sebek occasionally tried to make off with the sun, the eye of the world. The resulting eclipse set the priests of the sun god, Ra, at Heliopolis and Sebek's devotees at Crocodilopolis in a ceremonious frenzy to restore the sun and placate the gods.[7] Sebek was credited with causing blindness in man, a serious problem in ancient Egypt as in the modern Middle East where trachoma remains a plague. The Ebers Papyrus, a medical record of about 1550 B.C., lists ninety-two prescriptions for treating diseases of the eye; the chief ingredients of many are powdered lapis lazuli, crocodile dung and fat, and milk mixed with honey. These were applied with prayerful pleas to Horus, Isis, and Osirus, and after bandaging the patient's eyes, the physician would intone, "I have brought this thing and put it in place. The crocodile is now weak and powerless."

That most prescriptions were medically inert is reflected in a letter from a blind artist to his son: "News from the Painter Poi to his son, the Painter Pe-Rahotep: 'Do not abandon me! I am in distress. Do not cease to deplore me for I am in darkness. Bring me some honey for my eyes and some fat. I want my eyes and they are missing!' "[8]

Crocodile fat and dung, bile, blood, and scutes were all important items in Egypt's pharmacopoeia. After presenting the first clinical description of diabetes, the Ebers Papyrus advises the physician: "Thou shalt say to him, it is a decay of thy inside; and thou shalt prepare for him remedies against it: ground dragon's blood with flax seed boiled in oil and honey so that his thirst perishes and the decay of his insides may be expelled."[9]

By the first century A.D., when Pliny wrote his monumental *Natural History*, there was scarcely a disease of men, from pimples to impotence, which the crocodile could not cure. Romans carried stones from a crocodile's belly as charms against aching joints, bound crocodile teeth to their arms as aphrodisiacs, treated whooping cough in their children with doses of crocodile meat, and trustfully administered burned crocodile skin mixed with vinegar as an anesthetic to patients about to undergo surgery. Women used crocodile dung in a lint tampon as a contraceptive.

Pliny, who before dawn each morning waited on the emperor Vespasian, doubted that "the first turning joint in the chin of a dragon doth promise easy and favorable access to the presence of princes." And as an experienced advocate, he snorted, "Impostors all!" to the advice that "the fat growing about the heart of a dragon, lapped with the nerves and sinews of a red deer, assureth a man good success in suites of law." However, for his malarious friends, he recommended the use of "a crocodile's heart wrapped within a lock of wool which grew upon a black sheep and hath no other color meddled therewith, so that the said sheep were the first lamb that the dam weaned, to drive away the quartane agues."[10]

One of Pliny's prescriptions, the use of crocodile meat and oil "for an old cough," persisted in Europe for centuries. Settlers brought it to North America, and in the 1880s victims of tuberculosis with "one foot in the grave and a hand on the door knob" traveled to the southern states where they dosed themselves with alligator meat and rubbed their chests with alligator oil.[11] Until the discovery of streptomycin in 1944 the treatment was probably as effective as any other form of medication.

The Greek historian Herodotus, who traveled in Egypt in the fifth century B.C., observed that "some Egyptians regard the crocodile as sacred and some do not." He visited crocodile shrines along the shores of Lake Moeris near the ancient city of Crocodilopolis and described the sacred beasts as wearing ornaments of "molten stone" (glass) and gold in their ears and gold anklets bound around their forelegs. Priests and acolytes catered to their every whim, indulged them in luxury and, after they died, carefully mummified them and placed the bodies among the royal tombs in a labyrinth especially constructed for this purpose by the Saite Kings, 663–525 B.C.[12]

In the first century B.C., the Greek geographer Strabo, following in Herodotus's footsteps, recorded:

> We came to the city formerly called Crocodilopolis, for the inhabitants of the same worship the Crocodile. The animal is accounted sacred, and kept apart by himself in a lake; it is tame and gentle to the priest, and fed with bread, flesh, and wine which strangers who come to see it always present. Our host accompanied us to the lake and brought from the supper table a small cake, dressed meat and a small vessel containing a mixture of milk and honey. We found the animal lying on the edge of the lake. The priests went up to it; some of them opened its mouth, another put the cake into it, then the meat, and afterwards poured down the honey and milk. The animal then leaped into the lake and crossed to the other side. When another stranger arrived with his offering, the priests took it and running around the lake, caught the crocodile and gave him what was brought in the same manner as before.[13]

Strabo's contemporary Diodorus the Sicilian explained that the crocodile cult originated when the pharaoh Menas was set upon by his own dogs. He fled toward the lake with his hounds hot on his heels. When he arrived breathless at the shore, a friendly crocodile took him on its back and carried him to safety. The grateful king thereupon declared all crocodiles to be sacred and granted them divine honors.[14]

Variations on Egypt's crocodile cults persisted in parts of Africa into modern times, especially in Dahomey where the animals were bred in two pools near the royal palace at Savi. They were believed to harbor malignant spirits who harried travelers and required regular placation. Safe passage on lagoons and rivers might be purchased through the intervention of a special priestly caste who received offerings and prayers.[15]

As recently as 1967 Luis Marden, chief of the foreign editorial staff of the *National Geographic*, reported an active crocodile cult in Madagascar on the shores of Lac Sacre about 30 miles from Diego Suarez. Whenever a bride proves to be barren, or a villager falls ill, the people assemble on the lake shore with drums and wind instruments. Chanting and the loud clamour of the drums lure expectant crocodiles to the shore, where they are fed pieces of freshly slaughtered zebu and the votaries entreat them to intercede with Zanahary, a Malagasy deity, for the welfare of the village. The worshipers claim that long ago the site of the lake was a village where one day a thirsty pilgrim asked for water. Only one woman would help him. The traveler warned her to leave at once, and as soon as she had departed, a great flood submerged the community and all the residents were turned into crocodiles. Her descendants formed the cult.[16]

In many parts of Madagascar the crocodile is considered a supernatural beast to be placated by prayer and ceremony. The Malagasy people will not shake a spear over a stream for fear of annoying the animals and to throw dung or garbage into the water is considered a heinous offense. Natives wear a crocodile tooth as a charm against attack, and those in northern Madagascar believe that the spirits of chiefs pass into crocodiles after death.

On Lac Itasy, in the central part of the island, village elders make an annual proclamation to the crocodiles that the death of innocent men will be avenged. Should a swimmer be taken, the council convenes on the side of the lake, calling for the crocodiles to deliver up the guilty party and assuring them that no ill will is felt toward the other beasts.

Tangem-Boay, or ordeal by crocodile, is practiced by some Malagasy tribes. A man suspected of committing a crime is taken to a river where crocodiles abound. His judge stands on the river's edge and strikes the water three times, addressing the reptiles and begging them to show whether or not the prisoner is guilty. Bailiffs then push the prisoner into the water. He must swim to the opposite side and back. If he can do this successfully he is considered innocent and his accusers are fined four oxen. The swimmer gets two, the king gets one, and the court gets the last.[17]

Another ordeal by crocodile is practiced by the Shilluks at Malakal on the White Nile. A man accused of adultery has to swim the river. If he is taken by a crocodile he is presumed to be guilty; if he gets safely across, he must be innocent.[18]

Missionaries to Africa seem to have been especially intrigued by crocodile worship, and much of the material available today comes from their eye-witness accounts of ceremonies since lost. In 1911, the Reverend J. Roscoe, a British missionary, found a flourishing crocodile cult about the island of Danba in the Victoria Nyanza. A temple was dedicated to the reptiles, and the officiating priest pronounced oracles while wagging his head and snapping his jaws like a crocodile. He also presided over human sacrifices to appease the beasts and insure safety for boatmen and fishermen on the lake. Enemy tribesmen captured in raids and forays were brought to the lakeshore and laid out along the beach, their arms and legs broken to preclude their escape; then crocodiles came to drag them into the water.[19]

In 1857, David Livingstone, the famous Scottish explorer and missionary, reported that the men of the Ba-kuena, or Crocodile Clan, in South Africa held it hateful and unlucky to meet or even to see a crocodile, an accident likely to cause inflammation of the eyes. If the Ba-kuenas happened upon a crocodile they spat upon the ground and said, "*Boleo ki bo* [There is sin]," as a preventive charm. Nevertheless, they called the crocodile their father, celebrated it in their festivities, swore by it, and made incisions resembling a crocodile's mouth in the ears of their cattle in lieu of a brand. Livingstone said that if a man were bitten or splashed by a crocodile he was immediately banished from tribal society.[20]

The Wanyamwezi people, who in the 1880s lived about a hundred miles east of Lake Tanganyika, commonly believed that to kill a crocodile was an expedient way to commit suicide. No man could survive such an act.[21]

In one of the earliest missionary accounts of travel in Africa, Joanodos Santos, a Portuguese friar traveling to Mozambique in the 1670s, wrote: "God has endowed this country with a simple called Circirini growing on the margin of the River Sofala, with which the people who are acquainted with its virtues having rubbed themselves, the crocodiles dare not venture to bite them lest their teeth become soft as wax and without strength." Santos goes on to say that the natives test the herb by rubbing some on their heads and then checking their own teeth. If these seem softened, then they feel free to enter the water, protected from the crocodiles.[22]

In 1790, C. S. Sonnini, a French naturalist and traveler in Egypt, reported a variation on this idea. He claimed that Christians swam freely in the Nile because their faith protected them from crocodiles but that Moslems enjoyed no such immunity.[23]

Damsels and Dragons

The Nile crocodile has cast a long shadow in mythology and folklore over the cradle of Western civilization and thought. Carried by floods from the delta of the Nile and the mouths of some of the smaller North African rivers, crocodiles must have occasionally straggled through much of the Mediterranean basin in early historic times. A population was known to inhabit the marshes near Haifa, Israel, up to the early part of the present century. The originals of the fire-breathing dragons slain by classical and medieval heroes to rescue imperiled damsels were probably hard-puffing old Nile crocodiles that doubtless wished they were back in the papyrus swamps.

In March 1970, when we were enjoying lunch with Israeli friends in Jaffa, the ancient seaport which borders modern Tel Aviv on the south, our host pointed to a line of rocks well beyond the jetty and told us that here Andromeda was chained to wait for the sea monster when Perseus, enroute from gorgon killing in Greece, found and rescued her.

Perseus's exploit appeared in the Middle Ages disguised as the story of St. George and the dragon. According to the eighteenth-century English historian Edward Gibbon, George was really a scoundrel with all the lovable qualities of Al Capone. He began life as a fuller's thieving apprentice, graduated to selling rancid bacon to Constantius's army, and ended by plundering the citizens of Alexandria.[24] The story of how he rescued a maiden from the dragon seems to have been grafted on to his biography sometime in the twelfth century, possibly by enthusiastic Crusaders to the Holy Land.

Medieval naturalists recognized the dragon as a genuine contemporary animal, well described by travelers. At Aix, the fossilized head of an extinct saurian reptile was shown as the true head of Taraque, a dragon of the Rhone slain by St. Martha. In churches at Marseilles, Lyons, and Cimiers, skins of stuffed crocodiles were exhibited as remains of dragons.[25] About 1930, Major Flower saw a stuffed crocodile in a museum at St. Gall, Switzerland; it had been on exhibit since 1627.[26]

Stories of maidens sacrificed to dragons were largely allegorical in Europe but grim reality in Africa and Asia.

As the Nile waters rose in the delta, young virgins were thrown into the river at the time of the *khalig* (canal cutting) to placate the crocodiles and insure successful irrigation of the land. A similar custom may have existed in Israel and provided a factual basis for Andromeda's story. In Egypt, this practice persisted until A.D. 642, when the Moslem invader Amr Ibn-al-As put a stop to it. He authorized the substitution of wood or clay puppets for the living girls and these, called the "Bride Betrothed," have been used up to modern times.[27]

In eastern Asia, on the island of Timor, the Rajah of Kupang annually sacrificed a royal virgin to the saltwater crocodiles.[28] Kupang's royal family claimed to be descended from crocodiles. They sacrificed to the reptiles whenever a new prince was crowned, offering a pig with red bristles and a young girl, beautifully dressed, perfumed, and decked with fresh flowers. Guards escorted the girl to a cave near the water and set her on a sacred stone. Eventually a crocodile dragged her into the water. The people believed that he married her and if he did not find her virgin would return her to the cave.

Once, when the people of Cayeli in Buru, another East Indian island, were plagued by crocodiles, priests ascribed the menace to the passion which a crocodile prince felt for a certain girl. They compelled the girl's father to dress her as a bride and deliver her to the crocodiles.[29] In Korean folklore, the citizens of the city of Oop yearly assembled on the bank of the Han River to throw a lovely virgin into the water to appease its dragon spirit.[30]

In addition to maidens, other human victims have been fed to crocodiles. In old Burma, the royal palace was protected by moats filled with crocodiles who were regularly fed human beings, often war prisoners.[31] In the late sixteenth and seventeenth centuries, when Ceylon was a Portuguese possession, the saltwater crocodiles in swamps around Colombo provided the Portuguese garrison with a natural defense. Sinhala captives were thrown to the reptiles, which allegedly came to expect such feedings.[32] A similar custom continued in China, despite the disapproval of Confucius (551–479 B.C.), until the seventh century A.D.[33]

The Wani and the Son of Heaven

Japan is far beyond the range of any crocodile; nevertheless, the Japanese people recognize the animal and accord it a significant role in their folklore. They call it the *wani* and respect it as a mythological progenitress of Japan's imperial family.

In the deity age of Japan, soon after the islands had risen from the Pacific, a fair damsel whose name was Princess Blooming-Tree-Blossom became pregnant on her wedding night and gave birth to triplet sons. But her husband mocked her and, looking upon the children, laughed. "How can I, Sky-deity though I am, cause one to be pregnant in a single night? I suspect the children cannot be mine. They must be the children of a deity of the land."

The princess was sad and angry. She built a doorless hall and entered it with her babes, plastering it up behind her and vowing, "If these children be not the offspring of my husband, let us perish." Whereupon she set fire to the hall. When the fire was burning fiercely, the three baby boys sprang through the flames calling, "Where is our father?" And the Prince knew that they were his own sons.

He tried to placate his wife, who would not speak to him. Having proved her virtue, she left the babies in his care and climbed Mount Fuji where she became the patron deity of the volcano, hovering over the mountain's summit in a shining cloud and preventing from ascending any who were not pure at heart.

Meanwhile, the boys matured and Fire-Climax, the middle child, set out to seek his fortune. The remaining two, like brothers everywhere, coveted each other's life styles. Finally, Fire-Fade, the hunter, traded his bow and arrows for Fire-Glow's fishhook and each went off to try his luck. Fire-Glow found hunting wearisome, and Fire-Fade lost his brother's hook, so when they met again, they decided to trade back. Fire-Fade had forged five hundred fishhooks from his personal sword to give his brother but to no avail. Fire-Glow wanted his very own fishhook and none other. In despair, Fire-Fade sought the hook in the palace of the dragon king of the sea, where Princess Fruitful-Jewel fell in love with him. He wed the princess, got the fishhook back, and for three happy years enjoyed love and prosperity.

Princess Fruitful-Jewel became pregnant and she asked her husband to build a bringing-forth (parturition) hut for her on the sandy beach. He had nearly completed the house and was thatching it with cormorant feathers (a charm to ease her labor pains) when she appeared, riding on the back of a sea turtle. She entered the hut, saying, "When an outlander is about to give birth, she assumes the shape common to her native land. So now I am about to assume the shape of my people. Look not upon me, I pray you, while I am in travail!"

Fire-Fade could not restrain his curiosity. He peeked through the wall at the very moment she was delivered and saw his lovely wife transformed into a huge crocodile, writhing and twisting on the floor. Terrified, he fled.

Princess Fruitful-Jewel was angry and ashamed. She called to her husband, "Had you not disgraced me, I would have made land

and sea communicate, but your peeping was an outrageous act! With what now shall friendliness be knit together?"

She swaddled the babe in rushes and left it on the shore, closed the sea's boundary, and returned to her father's house.

Chastened, Fire-Fade busied himself with his son; choosing wet nurses, hot-water providers, chewers of boiled rice, and bathing women. From the circumstances of the birth, he gave the boy a melodious Japanese name which may be translated as "Prince-High-as-Sky's-Sun-Wave-Marge-Brave-Cormorant-Thatch-Making-to-Meet-Incompletely."

Deep in the sea in her father's house, Princess Fruitful-Jewel could not overcome the love in her heart. She sent her younger sister, Princess Jewel-Good, to nurture the babe.

When the young prince was grown, he wed his maternal aunt. She bore him five sons, the fourth of whom was Jimmu-Tenno, the legendary first emperor of Japan.[34] The present emperor, Hirohito, traces his ancestry in an unbroken line back 125 generations to the *wani*, Princess Fruitful-Jewel, daughter of Ryu-jin, the dragon king of the sea.[35]

Formerly the Japanese believed that to see a dragon in its entirety was to expose oneself to certain death, the punishment for looking upon too much divinity. This concept is said to be the reason that, in former times, Japanese emperors, lineal descendants of the dragon, were always hidden by a bamboo curtain from those they received in audience.[36]

Other Crocodile-Human Kinships

Kinship between humans and crocodiles has sometimes taken a bizarre form, especially in islands of the Indonesian Archipelago, including Timor, Java, Sumatra, and Ceram, where native women were once believed occasionally to copulate with crocodiles. The issue of such a union was always twins; one human and one crocodilian. The attending midwife would liberate the baby crocodile in the river, and thereafter the woman and her family placed offerings of food on the riverbank. When the human twin was sufficiently grown, he became responsible for seeing that his saurian sibling was fed, and in return, the crocodile was expected to protect the family by magic. On festive occasions, large boating parties of villagers would embark on the river with food and music, the people calling to their respective crocodile kinsmen to come and share the feast. When the crocodiles appeared, the celebrants tossed offerings of betel, tobacco, saffron, and rice.[37]

In the Papuan Gulf area, the Haura tribe, part of the Elema group, traces its ancestry to the first Haura woman who cohabited with a crocodile. Their children were the founding tribal chieftains.[38]

Australian blackfellows in the Gulf of Carpentaria area southeast of Arnhem Land say that in the "dreaming"—the time when myths were made—a huge crocodile roamed the country, his track being what is now Batten Creek. He deposited crocodile spirits wherever he paused to wallow and finally settled down at Wankilli, where there is a large pool with a stone in the middle of it. Men of the Crocodile Clan can increase the production of crocodiles by throwing sticks at this rock while chanting and singing. Since they depend on crocodiles for food, they frequently meet to have a sing-sing and induce the saurians to multiply.[39]

Kodal people of the Torres Straits claim the crocodile as their tribal totem. They wear crocodile scutes as badges of their superiority and sacrify their abdomens and chests with symbols representing the reptile. When a Kodal magician wishes to kill a man, he takes a crocodile's tooth, fills it with herbs, paints it red, and smears it with fat from a decaying human corpse. He then fashions a kind of slingshot from a young tree, using the tooth as a missile, and saying to it, "Now you go into that man, go into his heart!" He then releases the tension of the tree, the tooth shoots forward, and the man, theoretically, dies. The magician may also prepare a crocodile's tooth as a charm for burning an enemy's house, in which case he says: "Don't be lazy, you look very smart, you go and burn down that house."[40]

Some southeastern Asiatic people believed that human souls transmigrated into crocodiles. In the region of the Papuan Gulf, natives believed that the spirits of persons eaten by crocodiles were condemned to perpetual wandering and might often be seen near their home village. Sometimes they rested in the body of the reptile who ate them and appeared as a kind of fire in the animal's eyes, easily seen by close relatives. Such spirits delighted in harassing their former friends, sometimes throwing cold water on them as they slept. When a mischievious ghost got out of hand, his people lured the crocodile a long way up the river by trailing a bait from a canoe. Finally, they would abandon the canoe in the bush and slip away. The pursuing spirit would get hopelessly lost in the jungle and never find its way back to the village.[41]

The crocodile played an important part in puberty rites in the western Moluccas Islands south of the Philippines. Boys were initiated into manhood by being pushed through an opening shaped like a crocodile's jaws. The women bitterly mourned their fate, for they were considered dead. For a few days, the young men were instructed in the secrets of exclusively male society, after which the guardian priests returned to the village to comfort the women and announce that the devil crocodile had relented and restored the boys to life. Everyone rejoiced as the boys returned, now eligible to marry and establish families.[42]

In the western hemisphere, the alligator achieved the position of father-in-law to the sun and is credited by the Warrau, Akawai, and Makusi tribes of the Pomeroon River basin, British Guiana, with creating the first woman.

The Sun was a great fisherman and he kept his private ponds well stocked. Despite his care, each night many fish disappeared. Finally he appointed Alligator as night watchman, a lucky circumstance for Alligator, since he was the thief. Depredations continued until Sun caught Alligator carrying off a fine mess of fish. Angrily Sun slashed at the reptile with his cutlass, making a row of scales with every cut, while Alligator begged for his life, promising to give his daughter to Sun for a wife. Finally Sun agreed, but Alligator had no daughter. Surreptitiously he fashioned a lovely woman from a wild plum tree and offered her to Sun who found her anatomically incomplete and rejected her. Alligator called in the woodpecker, who obligingly pecked out the maiden's missing parts between her thighs, and Alligator drew a snake from her vagina. Sun gladly accepted her after these improvements and she bore him twin sons who were the progenitors of the Carib Indian tribes.[43] A Haitian Indian variation of this legend was reported by Ferdinand Columbus when he visited Haiti with his father, Christopher in 1497.[44]

In a puzzling coincidence, this legend is explicitly illustrated by two woodcarvings in the Volkerk Museum, Berlin. Each depicts a woman with her legs spread and a crocodilian pulling something from her vulva with its teeth. However, the figures are labeled as coming from Finch Harbour, New Guinea. Their possible connection with the mythology of South American Indian tribes provides a challenge to anthropologists.[45]

Crocodiles Remembered

As the seafaring peoples who eventually populated the myriad coral and volcanic islands of Polynesia fanned out from their ancestral homelands in southeastern Asia, they carried with them memories of the saltwater crocodiles, the great armored killers of estuaries and archipelagos. Crocodiles themselves are not known east or south of Fiji; crocodile legends crop up from New Zealand to Hawaii.

There are not and never had been any crocodilians in New Zealand. Yet Maori mythology teems with imaginary amphibious dragons who serve as boogiemen to keep children from straying from the village compounds at dusk. Maoris call these beasts the *taniwha*. They are fearsome, fire-breathing reptiles who make complicated tunnels and burrows where they lie in wait for the unwary and unwise.[46]

In the Hawaiian Islands, the royal family of Maui formerly worshiped Kihawahine, queen of the great lizards, who reputedly lived in a deep pool near the royal enclosure and watched after the welfare of the clan.[47] Hawaii has never harbored any native lizards; the islands have only small imported geckos and skinks that scamper up the walls of houses and hide themselves in the banana and pineapple fronds. Yet Hawaiian folk literature presents the most impressive of all dragon killers, Hiiaka.

This redoubtable female was the sister of Pele, goddess of volcanoes. The islands teemed with giant dragons, and Hiiaka and Pele agreed that they must be destroyed. While Pele stood guard in the crater of Mauna Loa, Hiiaka traveled from island to island, seeking out and killing the dragons. Finally she arrived at Kohala where the dragon god, Moo-Lau, held his court. Here began one of the great battles of Polynesian mythology.

Hiiaka threw off her leis and common clothes, took her lightning *pa-u* (skirt), and attacked Moo-Lau. He fought her in his dragon form, breathing fierce winds against her. He struck her with his writhing tail and tried to catch her between his powerful jaws. He coiled and twisted and swiftly whirled about, but she beat him back with her hands in which dwelt some of the divine power of volcanoes. She struck his great body with her magic skirt in which dwelt the power of lightning. Each struck with magic blows. Hiiaka grew tired. She sent out her spirit call. It broke through the clouds hanging over the crater of Mauna Loa and roused her sister Pele. A host of destructive forces swept down from the fire pit to aid Hiiaka. The slaughter of the dragon was soon accomplished, and Hiiaka continued her journey along the east coast of Hawaii until the land was free of dragons and the people could safely settle on the sandy beaches.[48]

Cook Islanders respected and feared Moko, a great lizard god with two hundred eyes, eight heads, and eight tails. The other chief deity of the Cook Islands was Rongo and the two gods were bitter rivals. Tonga Islands people tell a story of how Moko once stole a sacrifice from Rongo's altar. Rongo was enraged and vainly made repeated attempts to recover the offering from the watchful dragon god. Finally a great host of velvety butterflies and moths rallied round, picked up the sacrifice from under Moko's very nose, and carried it triumphantly back to Rongo, singing a war song as they flew along.[49]

Dragons of the Far East

The land and water dragons that permeate the culture of China, Korea, and Japan were doubtless based largely on the alligator of China, but behind this comparatively small and innocuous crocodilian glides the shadow of the great beast that lurked along the southern coasts of old China, the saltwater crocodile. Far beyond its natural range, it lives in legend.

China's dragon was originally a rain and fertility deity that eventually became a potent symbol for cosmic energy, merciful sovereign of all natural forces, and benevolent protector of Chinese culture. In the guise of a winged monster of gold and fire with lightning shooting from its talons, it personified the emperor.

Dragons proliferated in China. Sky dragons kept the heavens from falling; earth dragons traced the courses of rivers. Sentry dragons guarded hidden treasures. The great blue dragon of the east mounted the sky at the vernal equinox, rising on clouds of his own breath to unfold himself in the storm clouds and bring the spring rains. After harvest he descended and coiled in the depths of the sea.

Some Chinese dragons were lustful beasts who sought intercourse with other species. Coitus between a dragon and a cow produced a unicorn. An elephant resulted from a dragon's mating with a sow, and royal race horses were the offspring of a dragon and a mare.

Other dragons in China were the familiars of magicians. The legendary Ch'en Nan, who lived to be 1532 years old on a diet of dog meat and rice wine, carried his dragon in a gourd. The celebrated poet, Li Po, who drank heavily, is said to have ascended to heaven on a red dragon escorted by a blue fairy, a Chinese equivalent of pink elephants and mice.

Dragons of the storm could easily be controlled. They were terrified at the sight of a piece of silk dyed in five colors (a rainbow after the storm) and would disappear if touched with a piece of iron (lightning rod, which renders the charge harmless).

Fu Hsi, an early Chinese scholar (2852–2738 B.C.) who is credited with inventing the Oriental lyre and lute, is said to have enlisted the dragon's help in devising a system of written signs for the spoken word. Before Fu Hsi, records were kept with the help of knotted cords. His dragon is specifically described in Chinese literature. It had a flat head with two horns and long whiskers, a supple snake body with scales and spines along the back, and four legs with heavy claws. One day, while standing on the bank of the Yangtse River, Fu Hsi saw the dragon emerge from the muddy water with mystic signs clearly portrayed on its back. These Trigammata (primary symbols) which Fu copied from the dragon remained a basis for divination and magic in China for centuries.[50]

With the introduction of Buddhism in China, the native dragon was amalgamated with the Indian Naga kings (serpent deities), and this Sino-Buddhist dragon was transported to Japan. Engelbert Kaempfer, a German traveler who visited Japan in 1690, saw a dragon in the flesh, which he described as "a huge four-footed snake, scaley all over the body like a crocodile with sharp prickles along the back; the head beyond the rest monstrous and terrible."[51] The animal that Kaempfer saw was probably a crocodile imported from mainland Asia to serve as a patron deity for a Wani shrine.

In Japan the dragon became further diversified. It could be small as a silkworm or could fill the universe with its wings. It embodied both the male and female principles, was continually changing, accommodating itself to all surroundings like the everlasting cycles of life.[52]

Tatsmaki, the Japanese dragon of the sea, had a long watery tail. It was believed to live on the bottom of the ocean from which it periodically flew up into the air in the form of a waterspout.[53]

When the last wild crocodile has gone the way of the giant auk and only a few aging saurians remain to bask in remote temple pools and urban zoos, man will have vastly diminished his heritage from ancestors who saw in these reptiles the fabulous dragons who controlled the elements, founded dynasties, and fathered civilizations. For each day six dragons draw the sun across the heavens, while a seventh transports a philosopher to paradise.[54]

A Portfolio of Pictures

Three major crocodilians (*left to right*): American crocodile, gharial, American alligat

›te differences in snout shape and armor. (COURTESY ROSS ALLEN'S REPTILE INSTITUTE)

(*Above*) Head of the rare Cuban crocodile. The long fourth tooth in the lower jaw distinguishes crocodiles from alligators. (ANDREW KOUKOULIS)

(*Below*) The saltwater crocodile is perhaps the most dangerous of man-eaters. This half-grown specimen, photographed at Ross Allen's Reptile Institute, has unusually vivid markings.

Above) A Nile crocodile in a typical basking position, photographed at Ross Allen's Rep-
ile Institute. The species, immortalized in religion and folklore, once ranged throughout
nost of Africa and the southern Mediterranean area.

Below) Largest of the caimans is the black caiman, once very common in the Amazon ba-
in but now nearing extinction.

At Haleji Lake, a major source of Karachi's water supply, a sign warns of a dang
more fancied than real.

(*Above*) A sea turtle as pictured in Konrad Gesner's *Historia Animalium* (1551). (COURTESY INDIANA UNIVERSITY LILLY LIBRARY)

(*Below*) In terms of weight, the leatherback turtle is the largest living reptile. (AMERICAN MUSEUM OF NATURAL HISTORY)

(*Left*) The Moor Sahib, a large Indian crocodile, photographed at the Mangho Pir crocodile shrine near Karachi.

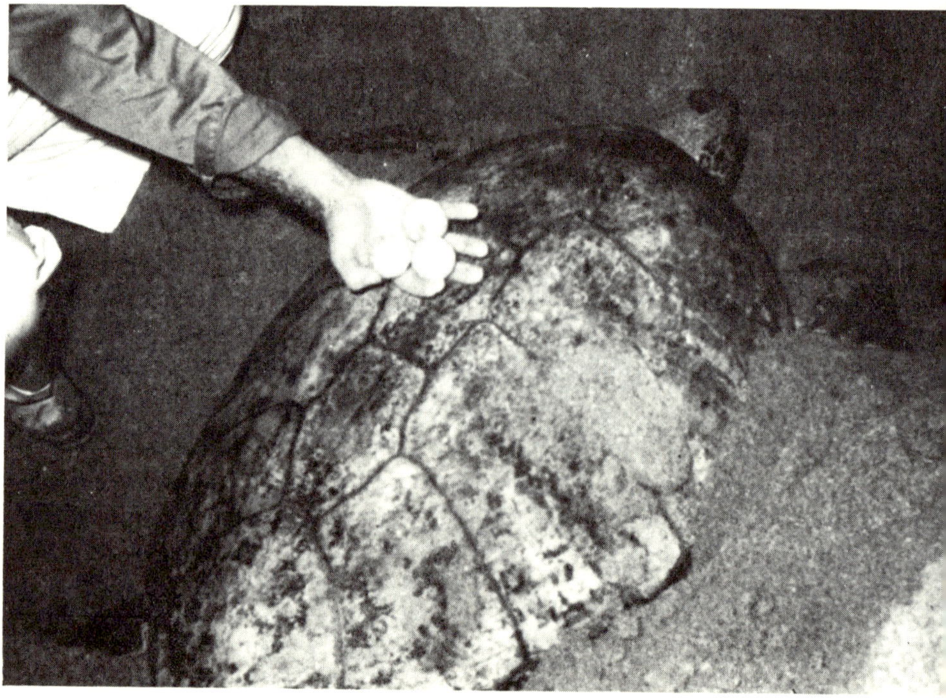

A baby green turtle heads for the sea.

(*Left, above and below*) A nesting sea turtle will actually lay her eggs in an outstretched hand. This green turtle, photographed at Hawke's Bay near Karachi, laid 120 eggs, some of which the authors ate for breakfast the next morning.

The lower shell of a female Pacific ridley caught at Hawke's Bay.

bove) Turtle hunters in the Indus valley with a 30-pound Ganges softshell that had
›e handled with caution.

·low) A narrow-headed softshell from the Indus near Hyderabad. This little-known
cies is one of the largest of the softshells.

(*Above*) Jesse Grammer of Grand Tower, Illinois, with a 170-pound alligator snapper, one of the largest of the freshwater turtles. (COURTESY JESSE GRAMMER AND PHILIP W. SMITH)

(*Right, above*) The common snapping turtle is found throughout most of the United States; this old fellow, though missing an eye, remains full of fight.

(*Right, below*) The Aldabra tortoise is the last surviving species of the giant island tortoises of the Indian Ocean.

›ne of the authors makes friends with a Galapagos giant tortoise. (ANDREW KOUKOULIS)

A Komodo dragon in its native habitat of Indonesia. Largest of living lizards, it is more massively built than other large monitors; the long, forked tongue is characteristic. (WALTER AUFFENBERG)

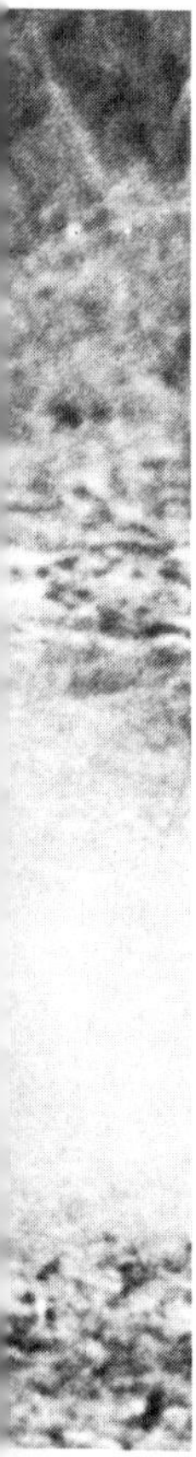

Although the species can be dangerous, this Komodo dragon in the National Zoological Park, Washington, D.C., seems on friendly terms with its keeper. (COURTESY NATIONAL ZOOLOGICAL PARK, SMITHSONIAN INSTITUTION; ROBERT G. TUCK, JR.)

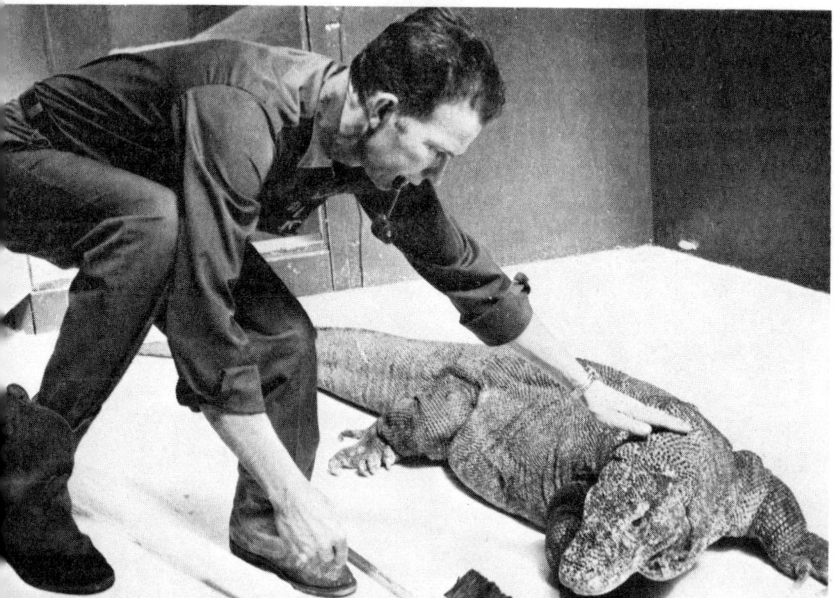

.eft) "The skinning of the aboma snake"—steel engraving of an anaconda from John nkerton, *Voyages and Travels* (1811). (COURTESY BUTLER UNIVERSITY LIBRARY)

ıe anaconda, a giant aquatic boa, is generally considered to be the largest snake; it ıabits many South American rivers. (ANDREW KOUKOULIS)

Indian pythons: (*left*) At the authors' back gate in Karachi, a Jogi woman offering a python for sale; (*left, below*) Neelo, a pet of the authors for several years and now a resident of the Philadelphia Zoo; (*below*) Omar, a Jogi snake charmer, displaying a freshly caught specimen.

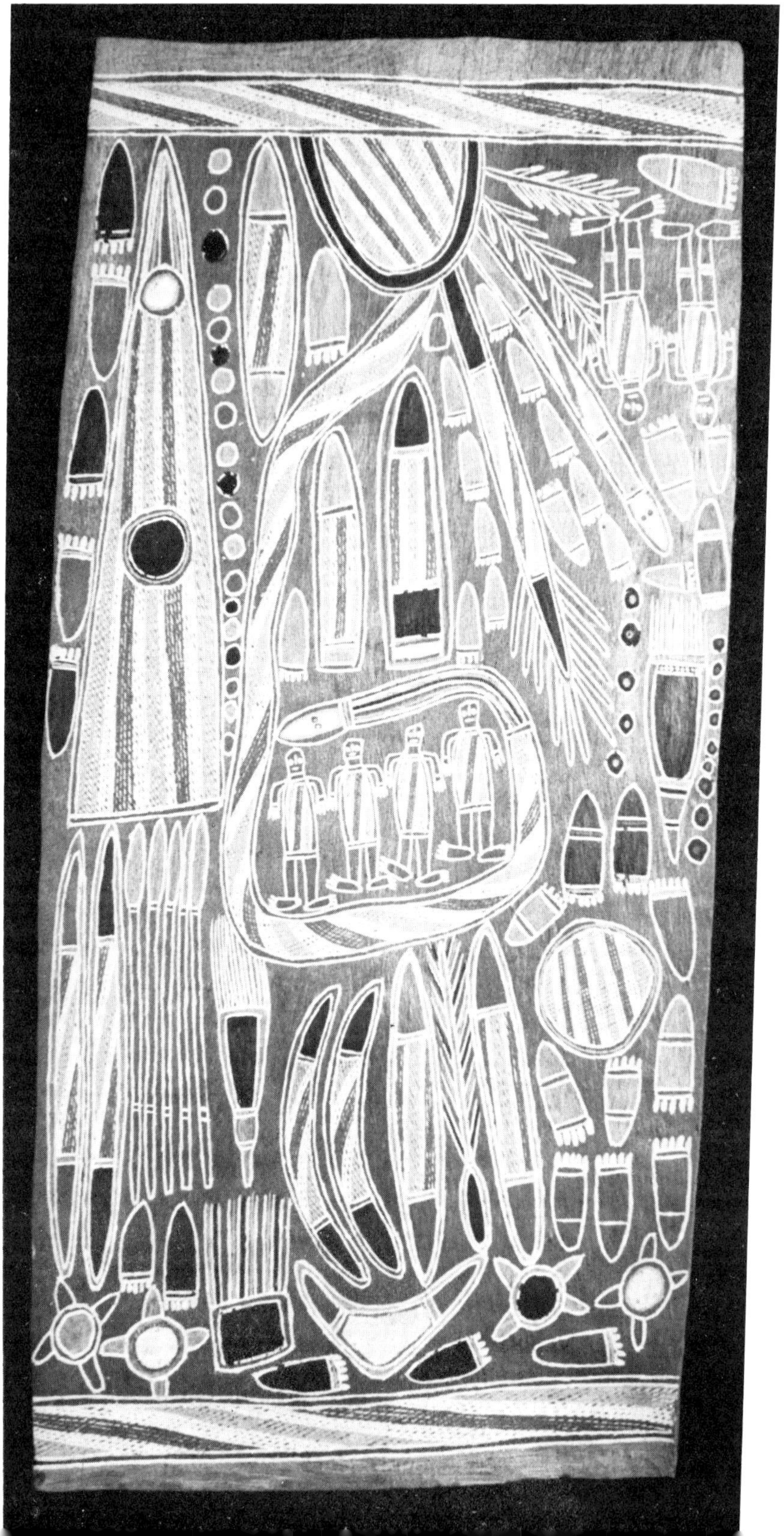

(*Right*) The constellation Serpentarius as depicted in *Hygini Augusti*, a 1549 work on astronomy. (COURTESY INDIANA UNIVERSITY LILLY LIBRARY)

(*Left*) Aborigine bark painting from Australia, illustrating the creation myth of the Liyagalawumirri tribe: the great Rainbow Python, Yulungurr, surrounds the Wawilak sisters and their children. (ALLAN ROBERTS)

oa constrictors have become popular as pets; the attractively marked variety shown ere is characteristic of northern South America.

A king cobra, largest of venomous snakes, is skillfully controlled by William E. Haast at the Miami Serpentarium. The hood of the king cobra is much narrower than that of the smaller Asian cobras. (CHRIS HANSEN)

6. Turtles around the World

If the last turtles had died off about the middle of the Jurassic and turtles were known only as fossils, there would be much speculation about how animals with such an odd and highly impractical type of bony armor managed to exist at all. They would be pointed out as a prime example of how overspecialization leads to speedy extinction. If turtles survived today only on some oceanic island or in some remote and ancient highland, they would have the living-fossil and zoological-curiosity status accorded the tuatara, the coelacanth, the lungfish, and the echidna. But although the turtles are the most ancient of living reptiles, they are as familiar to most of the world's peoples as birds, rats, and frogs. Some of the smaller species are even doing fairly well in modern urban America. As we drive through north Indianapolis during the warmer months of the year, we regularly see map turtles sunning along the water company's canal, and about once a year we may rescue a snapper or musk turtle that has forsaken the polluted waters of Fall Creek or White River for the even more perilous city streets.

The origin of the turtles is hidden in the early days of the age of reptiles, some 250 million years ago. All who have tried to trace their ancestry come eventually to *Eunotosaurus*, a little reptile from the Permian of South Africa, whose wide ribs seem to presage the formation of the turtle carapace. Its fossils are poorly preserved and incomplete, however, and it is impossible to say how it stands with respect to the true turtles. These appear rather abruptly in the

Triassic about 200 million years ago. Their skull is the skull of the old cotylosaurs, the stem reptiles from which all other reptiles arose, but they have managed to get their shoulder and pelvic girdles inside a bony box formed by the expanded ribs and breastbone. Once having accomplished this anatomical tour de force, they have been models of conservatism, whose mildly puzzled eyes have seen the dinosaurs and mastodons come and go, watched mushroom clouds rise over Einiwetok and Bikini, and seen rockets blast off for the moon. While showing a certain amount of physiological and anatomical versatility, they have remained unmistakably and unequivocally turtles.

The British, who have no chelonians (shelled reptiles) on their home islands, use the terms tortoise and turtle interchangeably for the whole group, although they seem to prefer the former. Americans do just the opposite. Nearly all the familiar native species are called turtles; tortoise has a faintly exotic connotation. Herpetologists consider the tortoises as a land-dwelling group coinciding with the family Testudinae; turtle is applied to shelled reptiles generally. Terrapin, is the name used in the United States for hardshelled, freshwater turtles, exclusive of the snapping turtles and their smaller relatives, the musk turtles. Beyond this point, the nomenclature gets complicated, especially in regions that have a rich and diverse turtle fauna. Turtle hunters in the southern states may talk of cooters, sliders, sawbacks, hacklebacks, snappers, softshells, and stink-jims, all of which names are quite explicit and definitive within a given area.

In India, along the middle Ganges where turtles are an important item of food, the turtle catchers recognize nine different kinds of turtles, which correspond well with the species known to herpetologists from that region. The distinction has practical implications, for some kinds of turtles may be eaten by even the Brahman caste, while others are forbidden to all but sweepers and persons of the lowest status. Interestingly enough, the river people of Sind in

West Pakistan seem to have no definitive names for the half-dozen or so quite distinctive turtles common in the lower Indus valley. At best they distinguish only between the hardshelled and softshelled species. This may be because these predominantly Moslem folk do not eat turtles and so have little incentive to distinguish one from another. However, our imperfect knowledge of their language may also have been a factor: one of my first attempts to obtain turtles from Sindhi fishermen netted me a nice collection of large aquatic snails.

While the snakes and lizards have many tiny species that can curl up comfortably on a silver dollar and weigh only a fraction of an ounce, there are none such among the more than 200 species of living turtles. The smallest turtles when adult have a shell length of about 3.5 inches and weigh about 4.5 ounces. Turtles, of course, run much more to weight and bulk than do snakes and lizards. A turtle with a shell length of 18 inches (measured in a straight line, not along the curve) can easily weigh 30 pounds, even if it is one of the more lightly built types such as the softshells. Any turtle with a shell length of 2 feet or more deserves consideration in the giant category. Most turtle families have such representatives, but the real giants are found among the sea turtles; the land tortoises, particularly the island races; the snapping turtles; some of the softshells; and a few Asian and South American river turtles. The sea turtles are considered in this chapter, the larger freshwater turtles in chapter 7, and the land tortoises in chapter 8.

Turtles of the Sea

The largest and one of the strangest of turtles is the seagoing leatherback or trunk turtle. Although it has the general shape of a turtle, its shell is not the usual massive bony structure but rather a mosaic of many small bones embedded in a thick skin. The form is beautifully streamlined, with seven low ridges on the upper side and five below. The outer, horny plates of most turtles are completely lacking. In a large adult animal with a shell length of about 5.5 feet, the enormous, winglike fore flippers have a spread of almost 9 feet. Unlike the flippers of other sea turtles, they have no claws.

Students of turtle anatomy and evolution find the leatherback something of an enigma. Many of the earlier workers believed that it represented an extremely primitive form, showing how the turtle shell developed from something like crocodilian dermal armor. A more generally accepted view today is that the leatherback diverged from the generalized sea-turtle stock about 65 million years ago and became adapted to a globe-girdling, pelagic existence. It is among the sea turtles as the giant albatross among the sea birds, a solitary wanderer with the whole of the temperate and tropical oceans as its home. No reptile in the world has a wider range. Although it is largely a tropical reptile, like all sea turtles, there are many records of its occurrence in temperate and even subarctic waters. Dr. Ilya S. Darevsky of the Zoological Institute of the USSR Academy of Sciences wrote us of a leatherback taken in 1967 a few kilometers off the Rybachii Peninsula between Murmansk and Vardo at the extreme northeast tip of Norway. This is in the Barnets Sea well north of the Arctic Circle. Within a few weeks in the summer of 1956 three leatherbacks were captured and a fourth sighted between the northern Shetland Islands and the Norwegian coast between Bergen and Målo.[1] Sherman Bleakney, a Canadian zoologist, compiled eighty-eight records of the leatherback turtle from coastal waters of New England and eastern Canada between 1824 and 1964 but mostly since 1889. He personally examined four specimens taken in Nova Scotia waters between 1961 and 1964.[2] In the Pacific, leatherbacks appear regularly off the coasts of Japan in two waves, one in summer centering about the northernmost island of Hokkaido, and one in winter in the Sea of Japan along the coast of Honshu. Along the Pacific coast of Canada, leatherbacks have been identified as far north as Sedgwick Bay in the Queen Charlotte Islands, which is very close to the southern tip of Alaska.[3] Less has been learned about the southern distribution of the leatherback; however, it is known from New Zealand, Cape of Good Hope, and the Isla de Chiloe which lies off the coast of Chile at about the same latitude as Tasmania. The cold currents that almost girdle the earth in the "roaring forties" may prevent the leatherbacks from making southern penetrations as striking as those they make to the north, or the paucity of records may only mean that there are fewer observers of sea life in the southern hemisphere.

Other species of sea turtles have been reported over nearly as extensive a global range as the leatherback, but there is an important difference. The other species are nearly always represented by immature individuals or weak or moribund adults caught in unfavorable currents and almost certainly doomed to perish in a hostile sea. Almost all the leatherbacks, on the other hand, are big, vigorous animals quite in command of the navigational situation and seemingly with a sense of purpose in their small chelonian brains. The seasonal, cyclic picture of their comings and goings suggests that voyages to colder waters are a regular part of their adult lives. Recent observations on nesting leatherbacks show that their temperature is slightly higher than that of two other sea turtles that nest with them and about 5°F. above that of the nearby sea water. The authors estimate that a leatherback in northern waters may be able to maintain a temperature 20° to 25° above that of the sea.[4]

Voyages to cooler latitudes may be related to the leatherback's diet, which seems to be quite as extraordinary as most other aspects of its life. There is now much evidence to indicate that these great reptiles, sometimes weighing more than half a ton and capable of tireless cruising and torpedo-like bursts of speed, feed largely on jellyfish and tunicates, or sea-squirts, creatures whose bodies are about 95 percent water. This seems somewhat like fueling a Cadillac with an eye-dropper. Few scientists have looked at the stomach contents of leatherbacks, but those who, like Bleakney, have done so have found jellyfish and some of the small fish and crustaceans that are regularly associated with them. Nothing else much has been reported except bits of seaweed that might have been swallowed accidentally or may make up a minor part of the diet.[5] Leatherbacks have been seen feeding on the Portuguese man-of-war, and the only large congregation of leatherbacks ever reported was associated with a dense concentration of the cabbage-head jellyfish, although the turtles were not actually observed feeding on the jellyfish.[6] The breeding season of the leatherback in Trinidad during February and March seems to coincide with a marked increase of jellyfish.[7] Another factor that may relate to a jellyfish diet is the peculiar backward-projecting flexible spines that line the leatherback's throat and esophagus. These are too slender and flexible to hold fish or other actively struggling prey but are just right for keeping a soft, slippery mass of material headed in the direction of the turtle's stomach. Of course, some jellyfish are huge animals despite their unsubstantial makeup. *Cyanea capillata*, a species that has been found in leatherbacks' stomachs, may measure 7 feet across the bell and trail its stinging tenacles 117 feet behind it. And jellyfish, under the influence of wind, currents, and other unknown factors, may form aggregations that carpet the sea's surface for large areas, presenting a considerable biomass to the predator that can harvest it. Although little is known of the formation and fate of such aggregations, they may provide a clue to the northern and southern wanderings of the leatherback.

Estimates of the size reached by the leatherback, like that of other giant reptiles, rest on a substratum of exaggeration and guesswork with the small amount of carefully collected data making a less impressive superstructure. Louis Agassiz (1807–1873), the Swiss-American naturalist whose work on the embryology of turtles was but one of several notable contributions to science, said that he had seen specimens of the leatherback "weighing over a ton."[8] It may be rude to question so outstanding an authority, but this statement sounds like an estimate and a generous one at that. A leatherback captured off San Diego in 1907 was said to have weighed 1,902.5 pounds, but the credibility of this estimate is suspect if the shell length of 5 feet 2 inches is correct. A turtle with a shell of that size should not weigh more than 1,000 pounds unless it had swallowed a few cannon balls. There are several acceptable records of specimens with shell lengths of 5.5 to 6 feet and weights of 1,000 to 1,500 pounds. Most of the carefully measured and weighed adult specimens taken in the last two decades have had shell lengths of 5 feet, give or take a few inches, and weights of 800 to 1,000 pounds. There seems to be no marked difference in size between the sexes.[9]

Four large communal nesting grounds of leatherbacks are known, the biggest on the coast of Malaya at Trengganu and the others in Zululand, Surinam, and Costa Rica. The southern coasts of Ceylon are another nesting site that has received considerable attention in zoological circles because of the excellent description of the nesting process given by Paul Deraniyagala.[10] A colony of about 100 leatherbacks nests annually at Matura Bay, Trinidad, and an equal number are estimated to nest on other beaches of the island.[11] Occasional independent-minded female leatherbacks haul out and lay their eggs on almost any beach in the tropics or subtropics from Flagler County, Florida, to Amami Oshima, an island in the Ryukyu chain between Japan and Taiwan. Sindhi fishermen told us that leatherbacks occasionally nested on islands near the mouth of the Indus, but we were unable to verify this. A large leatherback washed up, apparently dead or dying, on the green-turtle nesting beach at Hawke's Bay near Karachi. The nesting of the leatherback differs in no important respect from the nesting of other sea turtles, or, indeed, that of turtles in general.

Leatherback eggs seem to be quite as tasty and nutritious as other sea-turtle eggs and are just as sought after. At Trengganu, the adult turtles are not molested, and the taking of the eggs is under legal control. The species receives varying degrees of protection at its other nesting grounds.[12] The meat of the leatherback has the reputation of being unpalatable or even poisonous; however, several venturesome individuals, mostly herpetologists and marine biologists, have pronounced it, "like fine, aged beef, and it is extraordinarily tender. . . . The taste is like sirloin, but with a touch of the gaminess of venison . . . a real treat."[13] On Trinidad, about 20 to 30 percent of the nesting leatherbacks are killed by local villagers each year for meat and oil, although the turtles are legally protected.

Leaving aside the leatherback, the other sea turtles of the world make up a small, easily recognized family. Their comparatively flat shells are teardrop or shield shape when seen from above and have a fairly orthodox arrangement of horny plates. The limbs have become flippers but retain one or two well-developed claws on each fore flipper and sometimes vestiges of claws on the hind flippers. The neck is short, stocky, and less flexible than in other turtles. The head cannot be easily retracted, nor can the head and neck be used to right the turtle if it has been overturned or "turned turtle." This weakness has been exploited by turtle hunters, probably since the Stone Age.

The prototype and best-known member of the family is the Atlantic green turtle. The name is said to come from the greenish color of the fat. The adult turtles over much of their range are more brown than green, although those we saw nesting at Karachi were often olive to gray and looked appreciably different from the Caribbean adults we saw at the turtle farm on Grand Cayman. An average adult green turtle has a shell length of about 3 feet and weighs 250 to 300 pounds. There are old records of weights up to 850 pounds and comparatively recent and reliable records of shell lengths of 55.5 inches and weights up to 650 pounds.[14] Males have longer shells than females, but there is no great difference in weight.

At one time, green turtles were regularly found in the western Atlantic from New England to Mar del Plata, Argentina, which is about 225 airline miles south of Buenos Aires, and in the eastern Atlantic from the Gulf of Cadiz and the Mediterranean to the Cape of Good Hope. Moving eastward, the picture becomes confused. Green turtles much like those of the Atlantic are found in the Indian Ocean, but the Pacific green turtle is darker and somewhat different in habits. On lonely islands it sometimes crawls up on the beach to bask in the sun, something no other sea turtle does. It is found from the southern islands of Japan to southern Australia and from southern California to Isla de Chiloe, Chile. Whether the Atlantic and Pacific green turtles gradually blend or intergrade where they come in contact or whether each race maintains its identity is still uncertain, but it now seems clear that the flatback sea turtle of the Great Barrier Reef region of Australia is a distinct species. The *mestiza*, or yellow sea turtle, of the Galapagos is apparently a sterile mutant of the green turtle that may become extremely obese but never reproduces.[15] These mutants are known from other parts of the green turtle's range.

The loggerhead is another huge and wide-ranging sea turtle. Its head is wider than that of the green turtle, its shell is longer, and it may reach an even greater weight. Most specimens taken in recent years are about the same size as green turtles, but there are a few very large skulls in collections. The largest, slightly over 11 inches in width, is from Australia and is calculated to have come from a turtle weighing 1,192 pounds. The U. S. National Museum has a skull 10 inches wide that supposedly came from a turtle "killed in the Illinois River between Wauhillau and Stilwell, Oklahoma, in 1929."[16] This stream is a small tributary of the Arkansas River, and even the most dim-witted sea turtle that might have taken a wrong turn at New Orleans should have recognized its error before it got to Oklahoma. Be that as it may, this reptile, too, must have weighed a little over 1,000 pounds.

The loggerhead also has a tremendous geographic range. In the Atlantic it is known from Nova Scotia to the Shetland Islands, with an isolated record far to the north in the Barnets Sea. It ranges southward to Uruguay and the Cape of Good Hope. In the Indian Ocean it is widely distributed but not common, while in the Pacific it is reported from southern Australia to the northern part of the Sea of Japan and from southern California to Juan Fernandez Island off the coast of Chile. At least along the Atlantic coasts of the United States, it shows a tendency to swim up tidal streams into water that is only faintly brackish.

The ridley is a sea turtle superficially like a small loggerhead but with some important though rather technical differences. Ridleys are small as sea turtles go, having shell lengths of rarely more than 30 inches and a maximum weight of about 180 pounds. The range of the Atlantic ridley really centers about the Gulf of Mexico, although it occurs fairly regularly along the east coast of the United States to New England. Stragglers have showed up in the Azores and along the coasts of Europe to the British Isles. The Pacific ridley has a much wider range that includes much of the tropical and subtropical Pacific exclusive of the Australian region and the northern part of the Indian Ocean. Curiously enough, it also has well-established breeding populations on the coast of West Africa and on the northern coast of South America.[17]

The hawksbill gets its name from the beaked profile of the head. It differs from the other sea turtles in that the horny plates that cover the shell are unusually thick and in all but very old animals overlap extensively. These plates are the tortoiseshell used in the arts and crafts. The beautiful mottled pattern and color are not very evident in the living reptile. The hawksbill is also small, rarely exceeding a shell length of 30 inches and a weight of 100 pounds, though a 280-pound specimen is on record.

Although widely distributed, the hawksbill is believed to be more sedentary in its habits than the other sea turtles and less given to long journeys across the open sea. A strong homing instinct has been reported, however.[18] On the western side of the Atlantic, hawksbills are found from southern Brazil throughout the Caribbean and the Gulf of Mexico and along the Atlantic coast of the United States occasionally to New England. They occur in the Mediterranean and sporadically along the coasts of Europe to the British Isles but are rare or absent along the western coasts of Africa. The Indo-Pacific race of the species is known from the Red Sea to Madagascar and from southern Japan to southern Australia, as well as from many Pacific and Indonesian islands. Along the American coast, the range is from Baja California to Peru.

Migration

The biology of sea turtles, although important to maritime people over a good part of the world, has received little attention from biologists, with a few notable exceptions such as Archie F. Carr of the University of Florida. Not only is Dr. Carr's research of the highest quality, he has also produced some of the most readable prose written by any scientist in this century.

A distinctive feature of sea-turtle biology is the ability of the turtles to navigate vast distances of open ocean. A male leatherback, after a summer spent cruising off Norway, may go to Surinam to find a mate that may have swum from the mouth of the Rio de la Plata to meet him. Green turtles from the coast of Brazil lay their eggs on the beaches of Ascension, a tiny island about half way between South America and Africa. Almost all the Atlantic ridleys converge on a few miles of Tamaulipan coastline to breed. A loggerhead marked on the east coast of Florida was caught ten months later near the mouth of the Mississippi.

Periodic migration is not a common pattern in reptiles of today. Lizards, for all their speed and activity, usually stay in a small home range. If they make a long journey, it is likely to be a one-time thing. Admittedly most of them, including nearly every species whose habits are well known, are small animals. Some snakes, such as rattlers, travel up to a few miles every year between winter hibernating dens and summer hunting grounds. There are hints that some sea snakes make extensive migrations, but no one knows for sure. The hordes of caimans that once nested on islands near the mouth of the Amazon clearly had some cycle of migratory behavior of which the details will never be known. Crocodiles, particularly the saltwater crocodile, have made impressive sea voyages, but these all seem to have been individual efforts. But with the traditionally slow and conservative turtles a real migratory pattern is evident. It seems to be present in all sea turtles, although it is not well documented in the hawksbill or the little-known flatback. Some large river turtles, such as the common batagur of southeastern Asia and the Arrau river turtle of northern South America, also make periodic mass migrations to breeding grounds.

Although seafaring peoples who have hunted turtles for food have been aware of their migratory behavior for centuries, the best evidence comes from projects that involve tagging of large numbers of adult turtles. Tagging is not difficult particularly with females that come onto the beach to lay their eggs; one man has tagged as many as forty-five in a night. A large number must be tagged to get enough returns to give a picture of the species' migratory pattern, since many of the marked turtles are speared or netted on coasts where the population is semiliterate and the mail service uncertain, so the chance that a tag or the information on it will be speedily returned to the tagger's headquarters is not great. Turtle-tagging projects are now under way in several parts of the world, and information is slowly trickling in. As has been mentioned, the green turtles that nest on Ascension Island come from the coasts of Brazil; those that nest at Tortuguero on the Carribean side of Costa Rica wander as far as the Florida Keys and Isla de Margarita, which is close to Trinidad. Most females at Tortuguero nest every third year, but a few nest every other year, while around Australia's Great Barrier Reef and on the Turtle Islands in the South China Sea there are three- and four-year nesting cycles. Females show a strong tendency to return to the same nesting site each time.[19]

An unsolved problem is where the turtles go between hatching and sexual maturity. So far, no one has found a safe and practical way to mark baby turtles. Any conventional tag or brand that would be easily recognized on an adult turtle would be lethal or crippling to a hatchling. Tiny slugs of magnetized metal injected into the tissues are tolerated well by the baby turtles, but it is impracticable to locate these years later when the turtle has grown to a weight of some 300 pounds. Radioactive substances have been suggested as tags but would be a hazard to the turtle, to say nothing of those who might later eat it. A chemical marker of some sort would seem to be the answer, but no one has yet found a substance that is harmless, long-lasting, and easily detectable.

The physiological mechanisms that enable animals to locate small islands far from land, migrate across wide stretches of ocean, and find their way home across unfamiliar territory are not well understood. It is not hard to imagine how turtles find their way up and down rivers and along coastlines. The shore provides a changing visual panorama, while tributary streams, springs, changes in soil type, vegetation, and bottom-dwelling animal life must give every river and stretch of seacoast a distinctive chemical profile for an animal with the sensory equipment to read it. Finally, there are currents that can provide still other cues for orientation. There is every reason to think turtles can and do utilize all these sources of information. For navigation in the open ocean, however, the sea turtles must make use of "senses beyond the sense of man."

Green turtles, like many other animals, have a sun-compass sense that enables them to move in a general direction determined by the position of the sun in the sky. But for finding an island 5 miles wide separated from the mainland by 1,200 miles of empty sea the turtle needs something far more sophisticated. As Archie Carr puts it, "It is necessary to assume that all the star-map and earth-map sense that it has taken man so long to accumulate is stored up in the nervous systems of the navigating animals. . . . Is the whole Pacific a grid in the genes and brain of the golden plover? . . . Are the skies of the whole earth mapped by season and time of day in the mind of the Arctic tern?"[20] It is possible, as Carr suggests, that these patterns began to evolve when distances were not so great nor targets so small as they are today, but their presence is something to think about.

A newly advanced theory proposes that the open sea navigation of turtles may depend on a combination of chemoreception and sun-compass sense. After the hatchling turtle's first frenzy of swimming has taken it beyond the turbulent and hazardous surf, it is picked up by oceanic currents that bear it passively for many months and great distances. As it drifts, its brain is storing countless bits of chemical and physical data about each current and perhaps surprisingly unique to it. When the time comes for the turtle to return to the waters of its birth, it calls on this store of information and its compass sense to guide it back. "Evolutionary feasibility favors the idea: back-tracking an imprinted smell-stream to a tiny mid-ocean island seems a more evolvable adaptation than the Earth-map and star-almanac a migrant would need in order to correct offcourse aberrations by celestial bicoordinate navigation."[21]

The theory has not yet been successfully tested, and new experiments indicate additional complications. Adult green turtles released 37 kilometers (about 23 miles) off the Costa Rican coast took definite headings and crossed the shear line between the coastal water and the Gulf Stream without the least change in speed or direction. However, fixing a magnet to the plastron produced some degree of disorientation when the turtle was in deep water, indicating that the reptile may normally orient itself with the earth's magnetic field. The magnet did not have much effect in shallow water where the animal apparently used other cues.[22]

Nesting and Breeding

The migrations of turtles, like the migrations of most other animals, are intimately bound up with breeding and nesting. Although committed in nearly every other way to marine life, sea turtles cling to the basic chelonian custom of laying eggs in a hole dug in the ground. This is a deeply ingrained behavior pattern. The only turtles that do not follow it are the little North American musk turtles, which will leave their eggs almost anywhere—under a stranded boat, in an old mouse nest, or beneath a flat rock in a pasture—with shocking disregard for proper turtle mores. In spite of this, or perhaps because of it, they are surprisingly plentiful. Every other female turtle, from 1,000-pound leatherback to 6-ounce spotted turtle, digs a flask-shaped cavity with her hind feet, deposits 1 to 200 eggs in it, covers it up carefully, and goes her way without a backward look. In due time, the eggs hatch, and if the mother ever sees one of her offspring, she never knows it.

Pliny knew of sea turtles and their nesting habits, for he wrote, about A.D. 77, "There be found tortoises in the Indian sea so great that only one shell of them is sufficient for the roof of a dwelling house. . . . The female flieth from the male and will not abide to engender until such time as he pricke her behind and stick somewhat in her tail for running away from him so fast." He added that they commonly lay 100 eggs and beat down the nest with their breast. The females sit on the nest by night, and "they couvie a whole year before they hatch."[23] An English privateering captain, Woodes Rogers, came to a quite different conclusion from observations made on the Tres Marias Islands off the west coast of Mexico in 1709. He sent some of his men ashore to watch for a nesting turtle and ". . . suffer her to lay her eggs without disturbance and to take good notice of the time and place. Accordingly they did so and assured me they found the eggs addled in less than 12 hours and in about 12 hours more they had young ones in 'em, completely shaped and alive. Had we staid a little longer, I might have given myself and others a thorough satisfaction in this quick production of nature."[24] Undoubtedly Rogers's observations were made on a heavily used nesting beach, and his men inadvertently dug up nests with young nearly ready to hatch instead of the nest of the turtle they had observed laying her eggs.

We lived more than four years near one of the great sea-turtle nesting grounds of Asia, at Hawke's Bay, a curving beach of sand and shingle some 9 miles south and west of Karachi. Karachi is not noted for its night life. Teahouses are popular with the men who cluster about the radio and sip tar-dark tea, the movie theaters draw good crowds, and an occasional Asian or European dance troupe may give a performance at the big hotels. But one of earth's greatest and oldest shows plays regularly at Hawke's Bay.

We used to drive to the bay soon after sundown, stopping at the little guardhouse to be checked, as was mandatory for anyone leaving the ctiy after dark. Beyond this, a dusty road across the desert ended behind a line of dunes where Americans and Europeans had built a few dozen beach huts. It was a short clamber from the car to the top of a low sand ridge from which, if the moon were full, we could clearly see several hundred yards of beach with the waves slithering up the sand. On those nights when great blooms of flagellates such as *Noctiluca scintillans* lit the waters, each offshore breaker crashed in a shimmer of sudden light, and the receding waters left momentary threads of brilliance. In the phosphorescent water, huge turtle carapaces appeared as dark spots several yards from shore as the females approached. Once a turtle touched the shingle she would pause, thrust her head out fully, and survey the situation. This was a tense moment, for one rustle from us, and she would return to the waves, perhaps to try again farther down the beach. If we did not alarm her, eventually she would begin a slow journey, hitching herself across the sand toward the dunes, leaving a trail like the treads of a tank. Even then it was easy to alarm her, and she might, as hastily as possible, return to the surf. If undisturbed, she would continue up the slope, find a suitable spot, and begin to dig.

The first stage of the nesting process is the scooping out of a shallow basin about as deep as the female's shell is high. She does this by vigorous sweeps of her fore flippers that throw sand a considerable distance. It is common for a turtle to make two or three such hollows before getting down to the serious business of egg laying. The nest itself is dug only with the hind flippers, which are used alternately and most adroitly to make a smooth-sided hole about a foot in diameter and 15 to 18 inches deep. By this time the female's behavior is locked into a pattern that nothing short of dynamite will disturb. Often we have seen a turtle surrounded by a dozen people yattering away and knocking on her shell, while a hissing Petromax lantern threw a brilliant light and flash bulbs popped all around. The eggs, which are a little smaller than a tennis ball and have shells like tough paper, are extruded in bursts of six or eight with pauses between. In four nests whose entire construction we watched, we counted 81 to 125 eggs. The turtle will actually lay eggs in a watcher's hand, but everyone who watches one nesting has to convince himself of this and we were no exceptions. The turtle, however, never fails to cooperate. At the end of the laying she sometimes deposits a few tiny, misshapen eggs from the size of a marble to that of a pigeon egg. On one of our first trips to the beach, we found one of those little eggs lying all by itself. For some days we kept it in the hope that it was the egg of some rare snake, but it never hatched, and we eventually learned its true nature. After the last eggs are laid, the female fills the hole carefully, and as she leaves, she throws more sand to help hide the location of the nest. The whole process takes an hour and a half to two hours. During the whole nesting, the female's eyes flow with a steady drip of tears. "Weeping for her young who are destined to die," said a Pakistani friend. This weeping really goes on more or less continually, in the sea or on land, for the glands excrete excess salt that the turtles' kidneys are unable to handle. On the beach, the secretion has a secondary function of keeping the eyes moist and washed free of sand.

The main turtle-nesting season at Hawke's Bay is from late June through early November. During this season it was not unusual to count ten to twenty nesting turtles on the beach during a hunt of two to three hours. However, we saw turtle tracks on the beach during nearly every month and found hatchlings as early as July 21. Allowing for the standard two-month incubation period, those little turtles must have hatched from eggs laid in the latter part of May.

From September through most of November, baby turtles by the thousands erupt from the sands of Hawke's Bay. Most of them emerge at night, but we have seen them struggling toward the sea under a blazing afternoon sun. There is an appalling loss of both eggs and babies during the nesting season. The ubiquitous pariah dogs and their smaller cousins the jackals dig up many nests and later eat the hatchlings until their feces consist of little more than compacted masses of turtle plates. If they do not eat all the eggs, the crows, other birds, and monitor lizards finish the job. The sand crabs that swarm over the beach at night catch many hatchlings and tear them apart in a most gruesome manner. The turtles are to some extent their own population control, for we have seen a nesting female demolish the nest of an earlier arrival.

The one great turtle foe whose depredations are hardly felt at Hawke's Bay is man. Although Pakistanis from "up-country" sometimes disdainfully refer to their maritime compatriots as "turtle-egg eaters," we saw little evidence that either eggs or adult turtles were eaten in Karachi or in the coastal villages. Such is the power of religious proscription that literally tons of delicious meat and thousands of protein-rich eggs, freely available to a city of more than a million people, most of them chronically hungry, were never used. This prohibition is not universal in the Islamic world.

The presence of the city affects the turtle colony in more indirect ways. Although the pace is slower than in the Western world, the beach is building up, and some day there will be no room for the turtles. Moreover, the glow of the city lights in the night sky seems to disorient both the females and the hatchlings. Turtles apparently are guided to the sea largely by differences in light intensity between the sky over the sea and over the land. When man provides a distracting source of illumination, some of the turtles get confused. As mentioned, the beach at Hawke's Bay slopes upward to a line of low dunes. On the landward side of these is the beach road and beyond that are desert and a shallow, muddy mangrove swamp. Around the edge of the mangrove swamp we sometimes saw adult turtles that had died of heat and exhaustion or the attacks of dogs. One morning we arrived at the beach to see a party of Germans lifting a big turtle out of the road. They carried it over the dunes and into the surf with a good deal of singing and general *gemütlichkeit*. On this road we also saw our first hatchling turtles, flopping along rapidly but in the wrong direction.

About 95 percent of the turtles that nest at Hawke's Bay are green turtles. However, on one of our first trips we saw some plain, sooty gray hatchlings that were clearly not the same as the young greens with their shiny white plastrons and light-edged flippers and carapaces. Eventually we identified them as young of the Pacific ridley, we later saw a few female ridleys on the beach. We were able to watch the nesting of only one of these in its entirety. It was much like green-turtle nesting except that the female seemed quicker and more decisive in her movements and finished the entire process in less than an hour.

Green turtles usually mate in the waters off their nesting beaches. We watched two such unions during the early part of October. In both, the male had pulled himself up onto the female's carapace and was hanging on with the heavy claw on each front flipper. His shell was mostly out of water and his head depressed. The female's shell was visible only intermittently, her neck was fully extended, and she seemed to have trouble keeping her head out of water. The pairs appeared to be drifting with the current and showed little activity except for occasionally lifting a flipper in a gesture of bored and languid eroticism. I swam to within about 10 yards of one pair before they disappeared—as far as I could tell, without separating. Copulation is effected by the male's twisting his tail around and under that of the female until the vents are opposed. The chelonian penis is a large and bulky organ, and its insertion must require a good deal of time and cooperation. We do not know whether the eggs fertilized by the matings we saw in 1958 would be laid later that season or in 1961, assuming that a three-year cycle is normal for female green turtles in the Arabian Sea.

The migrations of the green-turtle population that breeds near Karachi are totally unknown. There seem to be some turtles around throughout the year. In late November and early December, after the nesting was essentially finished, we saw adult turtles of both sexes when we were snorkling at a reef a few miles west of Hawke's Bay. They were in water 5 to 10 feet deep, feeding on aquatic vegetation. On January 7, while we were sailing near the entrance of Karachi harbor, we glimpsed several big turtles among the sea weeds and one that was of hardly more than hatchling size. In April we saw turtles of 12 to 15 inches in shell length feeding on vegetation in coves west of Hawke's Bay. We assume they were green turtles, but they might have been ridleys or even hawksbills or loggerheads. There are many miles of lonely coastline between Karachi and the head of the Persian Gulf, and almost nothing is known of the sea-turtle populations that may nest on them. The only other rookery we know about is at Ras Jawani near the Iran-Pakistan border.[25] The northern part of the Arabian Sea, with its coasts sparsely inhabited by people whose customs prohibit their eating turtles or turtle eggs, must harbor one of the least-exploited sea-turtle populations in the world.

Turtles in Trouble

In most other parts of the world, the sea turtles are in trouble. Their eggs, especially the shell-less ones from the oviducts, are delicious eating, and in parts of Latin America they are reputed to be a powerful enhancer of virility as well. Beneath the plastron of every sea-turtle species except the leatherback is a gelatinous material known as calipee. It is the essential ingredient for epicurean turtle soups. In some places, nesting turtles are butchered for calipee alone; the rest of the carcass, including the unlaid eggs, is left for the vultures.

All sea turtles are edible. It is generally agreed that the green turtle furnishes the best meat and the loggerhead the poorest, but all can be good when properly prepared. Carr says of the part Chelonia played in the history of the Caribbean:

> All early activity in the new world tropics—exploration, colonization, buccaneering and the maneuverings of naval squadrons—was in some way dependent on the turtle. Salted or dried it everywhere fed the seaboard poor. It was at once a staple and a luxury—a slave ration and in soup and curries the pride of the menus of the big plantation houses. . . .
>
> Chelonia had all the qualities needed for a role in history. It was big, abundant, available, savory and remarkably tenacious of life. It was almost unique in being a marine herbivore—an air breathing vertebrate that grazed submarine beds of seed plants. It was easy to catch with simple equipment because its pasture lay under clear shallow water. . . . Chelonia's abundance expressed its straightforward ecology. It ate one kind of plant that spread continuously over great areas and knew no season. . . . The other turtles of these seas had gummed up their energy cycles by eating animals, which of course had their own complex problems and uncertainties. So the carnivorous turtles led a solitary, scattered life while the greens stayed in one place and grazed all day, only one link down the feeding chain from the sun itself. They grew fat and numerous and succulent, and in every way a blessing.
>
> Today the Atlantic green turtle can no longer be reckoned as a major asset. . . . One by one the famous old rookeries were destroyed. The first to go was Bermuda and next were the shores of the Greater Antilles. The Bahamas were blanked out not long after, and boats from there began to cross the Gulf Stream to abet the decimation in Florida. . . .
>
> One nesting ground stood apart from all the rest in its fecundity and stirred the wonder of all who saw it. This was the Cayman Islands. . . . The history of this fishery—its

> burgeoning and slow exhaustion, the wrecking of the rookery by heedless killing of the females that came ashore to lay, the plight of the people who had no other way to live and their tenacity in following the declining schools from one remote shore to another—is as extraordinary from the standpoint of human ecology as from that of resource depletion.[26]

Grand Cayman today is the site of a large green-turtle farm operated by Mariculture Ltd. Eggs from nesting grounds in Costa Rica and other sites in the Caribbean region are hatched here, and the young are raised in a series of circular plastic tanks supplied with running sea water. They grow rapidly on a diet of local turtle grass and the pelletized food used in catfish farms in the southern United States. After reaching a shell length of about 15 inches and a weight of 15 to 20 pounds, some are transferred to large concrete ponds on another part of the island. It takes a green turtle six to seven years to mature and the farm has not been operating long enough to develop its own breeding stock. In fact, there is some question whether turtles will mate and nest in the highly artificial surroundings of a turtle farm. If they will not, the main advantage of a farming operation will be in eliminating the high infant mortality of turtles in nature. The operation will still be dependent on eggs from wild turtles, and it will be essential to conserve adequate breeding populations. Whether the artificial propagation of sea turtles can be commercially successful remains to be seen. If it can, the turtles can go a long way toward alleviating the protein hunger of maritime peoples in the tropics as well as providing gourmet dishes elsewhere.

There are still a few wild turtles to be seen in thc waters around Grand Cayman. When we were diving there in the summer of 1971, one of our party spotted a little hawksbill resting on coral in 50 feet of water. It threaded its way gracefully through a group of about twenty divers and flapped off to vanish into the hazy blue.

The most spectacular sea-turtle nesting orgy is that of the Atlantic ridley on the coast of Tamaulipas in northeastern Mexico. On a day unpredictable by human reckoning, during April, May, or June, nearly every adult Atlantic ridley in the world heads for one spot on this 90-mile stretch of coastline. The result is called by the local people an *arribada*, the abrupt arrival of some tens of thousands of turtles bent on mating and laying their eggs in minimal time. While all other sea turtles nest at night, the Atlantic ridleys go about it in full daylight and in such numbers that a man could walk a mile on the shells of laying turtles and never touch his feet to sand. The nesting goes on at a frenetic pace, and so does the concomitant sexual activity, with males occasionally following females right onto the beach while trying to mate with them.

The scientific world did not learn of the *arribada* until 1961 and then only on the basis of a Mexican film made nearly 20 years previously. Until then the breeding habits of the Atlantic ridley had been a complete and perplexing mystery. Some scientists thought that the breeding grounds lay far to the north of other sea-turtle rookeries, perhaps on the shores of New England or the Azores. Others were almost ready to espouse the belief common among turtle fishermen that the ridley is a hybrid that never reproduces. And a few had even toyed with the idea that the Atlantic ridley might have utterly departed from normal turtle biology and become live-bearing. While the discovery of the mass nesting in Tamaulipas cleared up the major mystery, it raised new problems, such as how the turtles assemble, where the young and adults go afterward, and why such an odd reproductive pattern should have evolved in the first place.

Deplorably, this great nesting aggregation may have now vanished forever. In the past few decades, Mexican turtle hunters have been increasingly successful in arriving on the beaches at the same time as the turtles. When they did, they were utterly ruthless in gathering eggs and turtles, loading trains of forty or fifty burros with eggs and meat.[27] In 1967 Mexican fishermen took 10,540 tons of turtles. This catch, of course, includes species other than the Atlantic ridley.[28] A fair number of females of the latter species still nest from Padre Island on the south coast of Texas to San Andres Tuxtla in Veracruz, but the back of the reproductive effort has been broken. Whether newly passed conservation laws will protect the surviving remnant of the population has not yet been shown.

The Mexican government has instituted a program of protection for the nesting turtles and a tagging project to learn the extent of their migrations. Early returns on fewer than two dozen animals show that the general pattern of movement is along the Gulf Coast from the base of the Yucatan peninsula to the mouth of the Mississippi, with a single recapture between the Marquesas and Tortugas islands off the tip of Florida. This last turtle swam 2761 kilometers (ca. 1725 miles) between May 11 and December 10.[29]

Quite recently another but smaller mass nesting of ridleys was discovered on the coast of Surinam. Most of them come out at night, and their activity is not so frenzied as that of those that nest in Tamaulipas. But the real surprise is that they are not Atlantic ridleys but Pacific ridleys.[30] This is doubly incomprehensible because the Pacific ridley in all other parts of its far-flung range is the exact antithesis of its Atlantic relative. Instead of 10,000 turtles nesting on one beach, groups of a few Pacific ridleys nest on a thousand beaches from Michoacan to Mauretania. To add another enigmatic touch, there are several reports of large groups of Pacific ridleys seen far out at sea apparently headed for some unknown goal.[31] Turtles are old animals; somehow, though continents have drifted, land bridges broken and fused, and ice caps advanced and receded, chelonian migratory behavior may still reflect the times when it was first imprinted in brain-cell patterns that have changed less than the face of the globe.

The hawksbill has a triple price on its head. It is hunted for its eggs and meat like other sea turtles, but its real value comes from the horny plates, or laminae, that overlie its shell. These are much heavier than in other turtles, take a high polish, and are readily worked. The attractive mottled, flecked, or rayed pattern and the colors that range from clear pale yellow through warm shades of gold, amber, maroon, and olive to black have caught the eye of artists and craftsmen throughout much of the world. According to Pliny, the technique of cutting and laminating the shell for ornamentation of furniture originated with Carbilius Pollio, "a man very ingenious and inventive of such toies, serving to riot and superfluous expense."[32] The emperor Nero, one of the conspicuous consumers of his day, had a bathtub made of tortoiseshell. The Romans are said to have obtained their tortoiseshell from Egypt, and the Egyptians may have gotten most of their supply from India, although even today the hawksbill occurs sparingly in the Mediterranean and Red Sea. The artisans of China, India, Ceylon, and Japan have worked tortoiseshell for centuries, probably long before the Romans added it to their list of luxuries. Tortoiseshell is worked similarly to horn, but it is more brittle and less fibrous. The heat used for flattening and fusing the shell must be carefully controlled if the rich colors are to be preserved.

The finest tortoiseshell is said to come from Celebes and nearby islands of Indonesia, but other important sources are the Caribbean area, the Seychelles islands, Ceylon, Brazil, and New Guinea. In many places the turtle is killed before the plates are removed, but in some parts of the Caribbean the turtles are covered with burning grass or leaves, or boiling water is poured on the shell, until the plates can be pried loose. Turtles that survive are returned to the sea with the presumption that the tortoiseshell will be regenerated.[33] Probably most perish rapidly from shock or more slowly from infection, but some may recover. Loss of laminae is not necessarily fatal to turtles. When I was a boy, box-turtle races were popular for a short time in my home town. One of the best racers was "Moses," a large male box turtle that had lost most of his colorful shell plates, probably from having been caught in a grass fire. I don't know whether or not some dim memory of this narrow escape accounted for his extraordinary speed, but he was a strong competitor until he vanished, with most of our racing turtle stock, in the great Ohio River flood of 1937.

Sea-turtle meat of any species may cause serious poisoning at times. Dr. Bruce Halstead, an American marine biologist and physician who is probably the world's leading authority on poisoning and envenomation by marine animals, collected reports of ten incidents of turtle-meat poisoning between 1912 and 1965, as well as three earlier accounts. A total of 365 persons were involved, and 103 died. Nearly all the reports came from the Indo-Pacific region. In five instances, meat of the hawksbill turtle was involved; in one, that of the green turtle; in another, probably that of the leatherback; in the other six, the type of turtle was not identified. The rare and sporadic nature of the poisoning indicates that it is probably caused by some toxic organism that turtles occasionally eat. In this respect, and to some extent in its symptoms, it resembles ciguatera poisoning, which is caused by eating marine fish that have fed on toxic algae. This type of poisoning is a major problem in many tropical and subtropical regions. There is no way to identify poisonous turtle meat except by feeding some to a cat or dog, a custom followed in some places. Ordinary cooking does not seem to destroy the toxin.[34]

Disposition

Reports on the disposition and temperament of sea turtles vary, but everyone agrees that the green turtle is the best-tempered of the group. On land or in the water, it is completely inoffensive and will put up with almost any abuse with no gesture more hostile than an ineffectual waving of its flippers. At the other end of the spectrum, the Atlantic ridley displays an irascible temper when caught, snapping and thrashing its flippers. If turned upside down, as captured turtles frequently are, it will often die of exhaustion in a few hours, in marked contrast to the green turtle and most other species, which will survive for days in this position. The Pacific ridleys that we encountered at Hawke's Bay seemed quite as inoffensive as the green turtles, but we did not see many, nor did we interfere with them. Sailors from a U. S. warship who caught some Pacific ridleys from a migratory aggregation in open water found that the turtles could defend themselves effectively with the large claw on their fore flipper but made little attempt to bite.[35] Loggerheads apparently show a good deal of variation in temperament and can be formidable when thoroughly aroused. There is an account of five Connecticut fishermen who tried to recapture a 610-pound loggerhead that escaped from a pen at South Norwalk. After an attempt to spear the turtle failed, it allegedly turned on the boat and almost swamped it, bit off the oars, gashed one man's arm, knocked another unconscious, and retired from the fray uncontestably victorious.[36] The leatherback has been reported to defend itself almost as vigorously, and also to make a loud noise variously described as a snort, grunt, groan, or roar.

A recent newspaper report tells of two skin divers attacked by loggerhead turtles near Key Largo, Florida. One of them suffered a severe bite. Archie Carr said of the incident, "It would appear to me that the turtles were not trying to bite those divers—they were trying to mate with them. . . . This is the breeding season and the males get confused . . . what would just be a love-bite to a female turtle could result in horrendous injuries to a human being."[37] A male sea turtle in the grip of sexual passion appears to be singularly unperceptive and will force his attentions on anything that looks and moves vaguely like a female turtle. However, Julie Booth, an Australian naturalist who has used underwater gear to study turtles off the Great Barrier Reef, reports that the female turtle that is uninterested in sex assumes a characteristic position in the water that informs the male of her feelings and discourages his advances. This might be a useful gesture for divers to learn.

7. The Big Freshwater Turtles

The Softshells

The softshells are as odd a group of turtles as any that exist. Their shells are markedly flattened, and covering them is a leathery disc that extends well beyond the underlying bony structure. Head and neck are long and usually slender, and the nostrils open at the tip of a snorkel-like proboscis. The forelimbs are flipperlike, but not as markedly so as in the sea turtles. Every zoology textbook implies that scales or horny dermal plates are an essential attribute of a reptile, but in softshelled turtles these are hard to find. The shell, head, and limbs are covered with a smooth skin that often is intricately ornamented. Several of the Asian species, especially when young, have four large eyelike spots on the upper surface of the shell. In the peacock softshell these are made up of concentric rings of orange and dark purple on a dark-green background and produce a striking effect. Others have patterns that suggest the intricate calligraphy of Eastern art.

At least some of the softshells have a physiological adaptation unique among reptiles—namely, the ability to carry out pharyngeal respiration by pumping water in and out of the throat. In well-aerated waters they can stay submerged much longer than other turtles. They are, however, more vulnerable to toxic chemicals. We have been told by wildlife technicians that fish kills from the dumping of industrial waste into rivers are accompanied by high mortality among softshelled turtles but usually not among other turtle species. Stream pollution has virtually exterminated the Nile softshell in Israel.

Although the softshells have been around for at least 130 million years, they seem to represent a comparatively late turtle specialization. They are particularly well adapted to lying buried under silt or fine sand in water shallow enough that the snorkel can be poked to the surface for an occasional breath. They are undoubtedly the fastest swimmers of the nonmarine turtles, and some are fairly agile on land as well. They are particularly characteristic of slow-flowing rivers, but some species occur in lakes and ponds. One species is reported to enter salt water to a limited extent. The present-day headquarters of the group is southeastern Asia, with several species widely distributed in Africa and North America. Their fossils have been found in Europe, but they evidently never reached South America or Australia.

No softshelled turtle species is really small. The Indian flapshell turtles, which are the most primitive and probably the smallest members of the family, reach disc lengths of almost 11 inches and weigh up to 10 pounds. In trying to determine the largest size reached by softshells, one encounters the inevitable credibility gap. Dr. Nelly de Rooij, a Dutch herpetologist, in her classic work on the reptiles of the Indo-Australian Archipelago (1915–1917), credits Bibron's softshell with reaching a disc length of 1,290 by 740 millimeters (about 50.75 by 29.25 inches),[1] but it is not clear that she actually examined such a specimen herself. Clifford Pope examined a specimen of this turtle from Hainan Island that had a disc length of about 22 inches and weighed 42 pounds.[2] Local fishermen told him this was only about one-third the maximum size reached by the species. The British herpetologist Malcolm A. Smith, however, put the maximum disc length of this species at no more than about 2 feet.[3] B. L. Chaudhuri, an Indian naturalist, reported that the narrow-headed softshell, known as *sim* in Bengal, "grows to five feet in length of disc and weighs six to seven maunds" (about 480 to 560 pounds).[4] His information was apparently gleaned from local fishermen; no scientist has ever reported measuring or weighing a specimen approaching these dimensions. Smith considered this the largest of the Indian freshwater turtles and accepted for it a disc length of 800 millimeters (31.5 inches).[5] Although this turtle is prominent in Indian folklore, it is poorly known scientifically and may reach a larger size than any known member of its family. Other big Asian species with reported disc lengths of 70 centimeters (27.5 inches) or more are the Ganges softshell, the Indo-Chinese softshell, and the black softshell. The only serious contender for size honors outside Asia is the Nile softshell which, like the Nile crocodile, is found over a large part of Africa; it is reported to reach a disc length of about 3 feet. The largest of the specimens examined by Major Flower in Egypt and the Sudan had a disc length of 27.5 inches and weighed a little over 37 pounds.[6] This species also occurs in Israel, although it is now rare, and the biggest softshell we can vouch for personally is one that we saw in the Research Zoo of Tel Aviv University in 1970. According to Professor H. Mendelssohn of the zoology department of the university, this specimen had a disc length of 80 centimeters (31.5 inches) and weighed approximately 45 kilograms (about 100 pounds). The largest American species is the Florida softshell with a maximum disc length of about 18 inches. Among the American softshells, the females are considerably larger than the males. Professor Mendelssohn says the reverse is true of the Nile softshell. Information on the Asian softshells is inadequate.

Man has exterminated several species of giant turtles and endangered most of the others, but there is one species that apparently owes its survival to having become an object of religious veneration. This is the black softshell, which is known today only from a large pond in the city of Chittagong, East Pakistan. At this pond is the shrine of Sultan Byazid of Bastam, a Moslem saint who came to Chittagong from Persia in the eighteenth century, conducted devotional exercises there, and presumably died somewhere in the vicinity. According to one account, the turtles were once sinful men whom the saint metamorphosed into turtles as punishment for their wickedness. We have no information on the fate of this turtle shrine since the recent war and the creation of the new state of Bangladesh. Unless some direct hits by shells or bombs destroyed it, it will very likely survive. Local religious and cultural patterns in the Indian subcontinent are deeply ingrained and are little affected by the realignment of political boundaries.

The black softshell was first scientifically described in 1875. It is not clear whether at that time the turtles were restricted to the shrine or whether wild populations existed. Nelson Annandale, British zoologist and founder of the Zoological Survey of India, visited the shrine of Sultan Byazid in 1912, at which time it was still some 5 miles outside the city, but he knew of no other populations of the species, nor have subsequent workers turned up any. Annandale reported that the turtles were completely protected and would come to feed when called. The large ones were tamer than the smaller ones and would eat chicken from wooden skewers but usually refused bananas. They laid their eggs during the rainy season on a small hill in the enclosure. He estimated that the larger turtles were about 3 feet in disc length and extraordinarily massive in build. He was told that the oldest one was about 150 years old, but there was no way to verify this.[7]

We heard of the shrine during our stay in Pakistan but never had the opportunity to visit it. Guy Mountfort, a British ornithologist who visited Pakistan as a representative of the World Wildlife Fund, described it in 1966:

> The day being still young, we paid a visit to the Byazid Bustami shrine on the farther side of town. Here, in a turbid pool, hundreds of monstrous Sacred Black Mud Turtles were

> the centre of attraction. We entered the shrine down a narrow alley flanked by open stalls selling a gruesome array of raw offal, from which Black Kites and House Crows were snatching samples whenever the owners' backs were turned. Pilgrims to the shrine, armed with dripping pieces of cow's lung, were squatting at the edge of the pool feeding the turtles by hand. Some of the fungus-encrusted monsters must have weighed close on two hundred pounds and were obviously thriving on the diet.[8]

We believe that Mountfort's estimate of their weight is on the generous side. Nearly all observers tend to overestimate the weight of big turtles, just as they tend to overestimate the length of big snakes. Moreover, softshells, with their light bones, weigh less than other turtles.

The biology of the softshells is best known for the American species; there is much less information on the Asian and African ones. Four species occur more or less together along the lower Indus. The Ganges or Indian softshell occurs mostly in large streams and canals. In the spring, when the water is low, as many as twenty of these gray-green giants may be seen basking on the mudbanks or resting in the shallows with their snouts protruding. For a short time we had a captive specimen with a disc length of a little over 20 inches and a weight of 35 pounds. It lived in the tub in one of our bathrooms, which, because the plumbing was erratic even by Pakistani standards, was mostly given over to animals. It splashed about noisily and made disconcerting snaps at anything dangled near it, so we decided it was too much turtle for a family with small children and restored it to freedom. The huge narrow-headed softshell is far less plentiful, or at least harder to observe and collect, and we knew it only from sandy reaches along the main stream of the Indus near the villages of Sonda and Jerruck. The Indian flapshell is the most plentiful of the four species and the easiest to collect, for it often wanders about on land. It prefers lakes, marshes, and ditches to flowing water. The peacock softshell is a great rarity in the Indus and was unknown from that drainage until 1955, when Mustapha Konieczny obtained one at Jerruck. Ten years later Jerry Anderson got a beautifully colored baby specimen. All these species occur also in the Ganges, as do the other turtles of the Indus, the crocodilians, and most of the semi-aquatic snakes, thus indicating that these two river systems were probably connected in the past.

The American softshells are largely carnivorous, but this is not true of all the Asian species. The skull and jaw structure of the Ganges softshell and the narrow-headed softshell are quite different, the former being more adapted for chopping and crushing and the latter for cutting, and this strongly suggests a different diet. The throat and stomach of one Ganges softshell that I dissected were packed with neatly chopped pieces of aquatic plants, while the digestive tract of another contained about equal parts of plant and animal material. What the narrow-headed softshell eats I do not know, but the razor-edge jaws and the long, narrow head that can dart out with great speed look like adaptations for catching active prey such as fish. The sluggish flapshell is an omnivorous feeder.

Softshells are secretive about their sex life, and copulation has seldom been witnessed in any species. In the Chinese softshell, which is sometimes raised in ponds for food, this is said to occur at the surface of the water and to last only a few seconds although it may be frequently repeated.[9] The penes of the softshells that I have dissected bear long, soft, pointed projections that must have some function in mating, but what it is I do not know. Softshell eggs are spherical and in the Ganges softshell are a bit over an inch in diameter. The shells are hard but not as brittle as those of bird eggs. As is true of other aquatic turtles, the eggs are laid in holes excavated by the female with her hind feet in mud or sandbars. Average clutches of the bigger species contain twenty to thirty eggs. When the late Karl P. Schmidt and I were collecting along the Big Bend of the Rio Grande, we found an egg of the Texas softshell lying exposed on the sand. Naturally I assumed that a distinguished zoologist like my companion would preserve the egg with due documentation. Instead, he cracked it and ate it on the spot. Others throughout the world share Dr. Schmidt's fondness for softshell eggs, but the turtles spread their egg-laying over enough time and space to make the systematic looting of their nests generally unprofitable. However, along the upper Nile, Major Flower wrote that:

> As "Um Dirga" (mother of a shield) Trionyx is well known to the riverine Arabs . . . both the turtles and their eggs are very much prized as food by the Shagia Arabs: certain men make a specialty of turtle hunting and are very clever at it, knowing the turtle's habits. These men hunt at night when the turtles come ashore to lay their eggs.[10]

Adult softshells are eaten nearly everywhere they occur and apparently have been for a long time. In helping archaeologist friends sift through debris collected at Harrapan sites in the Indus valley (2000 to 1500 B.C.) we found easily recognizable pieces of the shells of the Ganges softshell and the Indian flapshell turtles that probably appeared on dinner tables in those long-vanished cities. Today, for the most part, the turtles are eaten locally by those who are skillful enough to catch them. Only the Chinese softshell, which has been introduced into Hawaii by the Oriental population of the islands, has been extensively exploited commercially.

Most species of softshells are bad-tempered and bite without hesitation, but solely as a defensive reaction. Some Indian naturalists, quoting local sources, have reported that the narrow-headed softshell "does not bite but disables its victim by blows [of its head], often injuring fishing boats by the impact thereof."[11] Since the vulnerable eyes and nostrils are very close to the tip of the snout in this turtle, the use of the head as a battering ram appears most unlikely. However, its great size and sharp, powerful jaws make it an animal worthy of respect. Pope reports that Bibron's softshell is normally sluggish but can strike with great speed. It keeps its jaws closed and gives a punch rather than a bite.[12] An animal caretaker at the Postgraduate Medical Centre in Karachi was bitten on the finger by a Ganges softshell weighing about 15 pounds. Although the finger was badly bruised, the skin was unbroken. One of our daughters carelessly held an even smaller softshell within range of her bare midriff. According to her account in a third-grade composition entitled "Our Zoo," "The bigger one tried to eat me. He bite me on the side. Oh how it hurt!"

Snapping Turtles

The giant of freshwater turtles is the alligator snapper. It is the largest living member of the snapping-turtle family, a small group entirely confined to the Americas. These turtles have massive heads with powerful jaws, thick necks, and the longest tails of any of the larger turtles. The head of the alligator snapper is enormous, about 7 inches wide in an average-sized adult, and the upper jaw terminates in a prominent hooked beak. The comparatively small but heavy shell has three high serrated ridges. As is often the case with aquatic turtles, it may become densely covered with algae, which help to conceal it as it lies on the bottom.

If a report of a 403-pound specimen of this turtle from the Neosho River in Kansas is discounted, the largest individual on record is one of 236 pounds that was exhibited in Chicago's Brookfield Zoo.[13] An individual of 219 pounds is also on record. An alligator snapper weighing about 160 pounds and measuring 54 inches from snout to tail tip was caught near Grand Tower, Illinois, in 1959 and was a major attraction in that little Mississippi River town during the 1960s, even being featured on post cards. It eventually reached a weight of 170 pounds. Records of alligator snappers weighing 100 to 150 pounds and with shell lengths of 21 to 25 inches are not particularly rare, but this is well over the average size. The turtles evidently mature at weights of 35 to 50 pounds and shell lengths of about 14 to 16 inches. Males are larger than females, but for nearly all the biggest specimens, the sex is not recorded. Growth to sexual maturity requires 11 to 13 years.[14]

The alligator snapper is characteristically a reptile of the rivers of the mid-South, ranging east to northern Florida and the Okefenokee Swamp and west to eastern Texas and southeastern Kansas. In the Mississippi it has been reported north to Rock Island, Illinois, which is almost as far north as Chicago, although the northernmost records are probably based on stray adults. The northernmost known breeding population seems to be in southern Illinois.[15]

Since the alligator snapper is nowhere very common, its habits are not well known. Although it prefers large, slow-flowing streams, it is also found in large swamps, ox-bow lakes, and canals. The adult turtles seem to be extraordinarily sluggish animals that usually live in water 5 to 6 feet deep beneath an undercut bank or in the shelter of a huge submerged log. Here they may rest almost motionless for hours or possibly days except for ponderously paddling to the surface periodically to replenish their air supply. They may also get some dissolved oxygen from the water by pharyngeal respiration but are not so efficient at this as are the softshells. Some observers report that alligator snappers are more active at night.

Despite their over-all sluggishness, alligator snappers do occasionally move considerable distances. One observed in an Oklahoma stream for 3 years traveled upstream approximately 6 miles a year.[16] They are also probably very long-lived and, when well grown, virtually without natural enemies. A specimen in the Philadelphia Zoo lived 58 years and 8 months. It was well grown when the zoo obtained it, and its death was accidental rather than from the infirmities of old age. Another, still alive in the zoo, was acquired from the now defunct Philadelphia Aquarium, where it lived from about 1916 to 1962.[17]

The alligator snapper's feeding habits are geared to a sedentary life. It is the only turtle that fishes for its dinner. Attached to the floor of its lower jaw is a wormlike, fleshy structure free at each end. When the turtle is fishing, the lure becomes pinkish from suffusion with blood, and its ends squirm and wiggle. Sooner or later an unlucky fish or smaller turtle investigates the bait, the great jaws snap shut, and the snapper has its meal. Alligator snappers also feed on snails and mussels and evidently do some scavenging, for they have been taken on hooks baited with meat. The big turtle exhibited at Grand Tower ate 70 to 110 pounds of meat during the warm months of the year, a very small intake of food for an animal of such size. We once kept a hatchling alligator snapper in the same aquarium with a common snapping turtle of about the same size. It was soon apparent that the more active and aggressive common snapper was getting a good 80 percent of the food and keeping its companion in a state of semistarvation. In the Yazoo River, where both were found, the situation would undoubtedly be different, for the food supply would be richer and more diversified, and each species could select a niche where it might thrive. Nevertheless, the alligator snapper seems to have sacrificed a good deal of its adaptability in exchange for its great bulk.

Alligator snappers are powerful animals, and most early observers who wrote of them were chiefly impressed with their ability to walk about with a man standing on their back and to bite broom handles in half. The biting ability of this turtle has been to some extent downgraded by Ross Allen and Wilfred Neill, who wrote:

> The savage bite inflicted by *Macrochelys* has been often mentioned, in fact, over-emphasized. Particularly irksome is the repeated story that a large *Macrochelys* can bite sections from a broom handle. . . . We find that specimens of 35 to 40 pounds weight are scarcely capable of biting an ordinary pencil in two. . . . Specimens of 90 to 100 pounds weight merely dent an ordinary broom handle.

However, in the same article they tell how a freshly captured turtle bit a chunk out of the wooden gunwale of a canoe. They also say, ". . . but when given some yielding substance such as a human hand, they retain their grip, pulling and chewing to inflict a painful wound."[18] These authors give no details of human injuries caused by the big turtles, nor have we been able to find any. This no doubt reflects the comparative rarity and inoffensive disposition of the reptiles.

The alligator snapper is not considered a particularly good food turtle, but it is occasionally eaten. The last specimen reliably reported from Indiana ended up in the soup kettle, although its shell was saved. This was in the mid-1930s when finding 50 pounds or so of edible turtle could make a real difference in a poor man's budget.

Although the New York sewers contain no alligators, some ponds and streams in the metropolitan New York area probably still harbor populations of the common snapping turtle. This is the largest reptile native to the northern United States. Not only does it survive in densely populated and industrialized areas, it actually seems to thrive there, although the degrees of pollution now being reached in some waters exceed even its tolerance. This turtle inhabits all of the United States and southern Canada east of the Rockies. Other very similar snapping turtles are found from eastern Mexico to Ecuador. Curiously enough, these tropical snapping turtles are uncommon, and almost nothing is known of their habits.

Apparently the adaptations that make snapping turtles so successful over most of the United States are less satisfactory in the tropics.

Most adult snapping turtles have shells 9 to 13 inches long and weights of 8 to 20 pounds. The biggest one we ever captured and weighed was a mere 22 pounds. This turtle, incidentally, was seen lying on the bottom of a stream under a half-inch of ice and about 15 inches of very cold water. I once caught what I believe was a considerably bigger one near the southwestern tip of Indiana. The task of lugging this monster through a mile or so of cypress and buttonbush swamp proved so arduous that I eventually gave it up. Forty-pounders are not unknown in the Midwest, although they are rare enough that they usually rate space in the local papers when captured. The largest wild-caught turtle seems to be a 68-pound specimen captured on the Valentine National Wildlife Refuge in Nebraska.[19] There is a record of a turtle fattened in captivity to a weight of 85 pounds.[20] The shell of snapping turtles is relatively small, not much over 15 inches even in specimens of 30 pounds or more. A great deal of the weight of large turtles is in the neck, tail, and limbs.

Snapping turtles seem to spend a great deal of their time in water shallow enough that they can stretch their necks and poke the tip of their snouts to the surface for air while their bodies rest on the bottom. They swim fairly rapidly, usually paddling along just above the bottom. In spite of their basically aquatic nature, they sometimes make extensive overland journeys and show up in unexpected places. I saw one emerge from beneath a dilapidated building and plod nonchalantly down the main street of a small Indiana town. Many are killed on highways. Our observations in the Midwest indicate this peripatetic urge seems strongest in the late spring and affects turtles of both sexes and all ages. It does not seem to be related to the reproductive cycle or to the drying of ponds and streams.

Snapping turtles may nest during almost any time of the year that the air temperature does not fall below 45° Fahrenheit but they usually do so in early summer. The eggs are spherical and a little smaller than a table-tennis ball. There are usually twenty to thirty in a clutch, but a big female may lay up to eighty-six. They hatch 70 to 120 days later.[21] If the eggs are laid late in the season, the young may remain in their underground nest all winter.

Snappers are voracious and indiscriminate feeders, taking almost any animal they can capture as well as considerable amounts of aquatic vegetation. They are also efficient scavengers. An Indian who lived in northern Indiana took a macabre advantage of this trait. He used a snapping turtle tethered by a long wire to locate the bodies of drowned persons. It is said that he was often successful when other methods had failed.[22] Sportsmen regard the snapping turtle with considerable disfavor because of the allegedly high toll it takes among game fish, young waterfowl, and young muskrats. There is some truth in this, although little evidence that the damage is serious in a more or less undisturbed environment. However, anyone who plans to raise ducks in a farm pond would be well advised to eliminate all large snapping turtles beforehand.

The snapper itself is eaten more frequently than any other North American turtle. Commercial turtle hunters use two general methods, trapping and noodling. A turtle trap is much like a large minnow trap, being made of netting stretched between a series of hoops with a funnel-like opening at one or both ends. For snappers the bait is usually meat or fish. The noodler wades or poles a boat through good turtle habitat and locates the reptiles by feeling with his hands or probing with an iron rod. When a turtle is located, it is hauled out by the tail. Noodlers sometimes worry about being bitten by snakes or muskrats but not by turtles, for the snapper's only impulse as long as it is under water is to escape. Some noodlers work during the winter when the normally solitary snappers congregate in muskrat burrows and under logs for hibernation.

Catches of as much as 5 tons of turtles per season have been reported from the Mississippi valley.[23] Snapper reaches its zenith as an epicurean dish in the Philadelphia area. It is more plebeian fare in the Midwest, where turtle soup and fried turtle were regularly on the menu in taverns of river towns during the 1930s.

Although snappers catch their food under water and sometimes fight among themselves while submerged, we have never heard of a well-authenticated instance of a turtle's biting anyone who was wading, swimming, or playing in the water. Most alleged turtle bites are much more likely to have been the result of stepping on broken bottles or sharp pieces of metal. Away from the water, a big snapper is an animal to be reckoned with, for it can strike with great speed and power, and its beaked jaws inflict a painful wound. The few authenticated bites that have come to our attention were sustained by persons catching or handling turtles out of the water. A physiology student was painfully nipped by the severed head of a turtle he was preparing for an experiment. The turtle's tenacity is embodied in the widely held belief that it "won't let go 'til it thunders" or, in Mexico, "won't let go until an ass brays."

Until about 1960, we had monsters in Indiana every summer. Usually they were giant snakes "as big around as a nail keg" or that "left a track as wide as a truck tire." Sometimes they were four-footed beasts that roared, screamed, left huge pug marks in the dust, or made off with dogs or livestock. But the monster of 1949 was a turtle. The story really began in the summer of 1948 when Gale Harris, a farmer living near the town of Churubusco in the northeastern part of the state, looked from the roof of his barn over the waters of nearby Fulk's Lake and saw "what looked like a submarine on top of the water. When I looked closer I saw it was a turtle." There were no further sightings until about the first of March, when the turtle was relocated and described as being "as big as a dining room table," covered with moss, and estimated as weighing 500 pounds. The first serious attempt to catch the turtle was made March 10, using an improvised dredge that the reptile easily eluded. On March 12, the monster was located on the bottom in about 20 feet of water and surrounded with a ring of stakes. The next day thousands of curious sightseers embroiled themselves in Whitley County's greatest traffic jam, and the turtle rose to the occasion by breaking out of the underwater stockade. By this time, catching the turtle had become a matter of community pride, and a professional diver from Fort Wayne was engaged to attach a net to the behemoth while an automobile wrecker was prepared to drag it ashore. The turtle took this round in a breeze when the diver's suit sprung a leak.

By mid-April, interest had flagged, although every large turtle caught in northern Indiana that season, and some that weren't, rated at least a local story as the fabulous "Beast of 'Busco."[24]

I didn't get personally involved in the hunt until about midsummer. My strategy was simple. If there was a big turtle in the lake, it had standard snapping-turtle instincts and physiology and was probably spending most of its time in comparatively shallow water. I would use a face mask (a comparatively little known gadget in Indiana in 1949) to locate the turtle. I had no illusions about my ability to drag even a 30- or 40-pound reptile out of the lake single-handed, but I hoped to dive down and hook a float on a strong, thin line to some part of the creature's anatomy, so it could be tracked and eventually captured. When I got to Fulk's Lake, I found its seven acres dotted with a formidable assortment of traps, some of which looked capable of catching and holding a small submarine. Mr. Harris, the diver, and a couple of other never-say-die turtle hunters were scanning the water without too much enthusiasm. The diver's rubber suit and metal helmet resting on a rough bench gave a slightly unreal touch to the scene. It took about 15 minutes for me to realize that I wasn't going to be the hero of Churubusco either. The lake was colder than I'd counted on, the water was the color of strong, murky coffee, and there were enough water weeds to hide an aircraft carrier. I spent the rest of the day swapping snake and turtle stories with the other hunters.

By September, Mr. Harris was a frustrated and unhappy man. His farm looked as though the Sixth Armored Division had been holding maneuvers there, and the turtle was still at large. In a last desperate move, he tried to drain the lake. This was ill advised, for Fulk's Lake, although small, was an old glacial pothole and very deep. The drainage efforts only created a wide margin of almost bottomless muck in which two turtle hunters became entrapped and nearly lost their lives. That was the end of the search for the "Beast of 'Busco." There are those who believe it still lives in Fulk's Lake or one of the other lakes in the area, but no one has seen it for a long time, and today inquiries about giant turtles are usually met with looks of blank incomprehension or wry smiles.

Batagurs and Other River Turtles

The most familiar turtles of the United States—the painted turtles, map turtles, box turtles, and red-eared turtles—all belong to the large family Emydidae, whose members dominate the freshwater turtle fauna of the northern hemisphere. The largest members of the family are found in southeastern Asia and reach sizes that warrant their inclusion among the giant turtles. Although these turtles are not particularly closely related to one another, for the sake of convenience we will refer to them as batagurs. The common batagur, which is found over a large area of India, East Pakistan, Burma, Thailand, and Malaya, reaches a shell length of about 2 feet and a weight of 50 to 60 pounds. It lives in deep, slow-flowing rivers and canals, often close to the seacoast. This is one of the few turtles that show a color change associated with breeding.

At this time, the head of the male is jet black, the neck and forelimbs crimson, and the area around the nostrils pale blue. In the estuary of the Irrawady River in Burma the turtles nest from January to early March. At night the females assemble in herds of 100 to 500 and come ashore to dig their nests in the sand. In contrast to the state of complete obliviousness observed in nesting sea turtles, female batagurs are very shy and cannot be approached. Nevertheless, by probing with sticks the natives locate many nests and feast upon the eggs. They also catch the adult turtles in basket traps baited with leaves of the *thame* tree.[25] Burmese Buddhists adorn the carapaces of these turtles with gold leaf and release them with great ceremony.[26] In the British Museum we saw what appeared to be a life-sized carving of a batagur beautifully done in jade.

The roofed batagur, or *sal*, is a large species whose shell is ridged rather than rounded. It also inhabits rivers of eastern India and Burma. Both the adult turtles and their eggs are eagerly eaten by some of the fisherfolk.

The larger islands of Indonesia are home for two other large species, the painted batagur and the Borneo batagur. De Rooij attributes a shell length of about 30 inches to both these species,[27] which, if correct, would indicate weights of 75 to 100 pounds. The Borneo batagur seems to be an exclusively freshwater species with a preference for animal food, while the painted batagur is largely restricted to brackish water and is a plant-eater.[28]

Largest of the South American river turtles is the Arrau turtle of the Amazon and Orinoco river systems. It reaches a shell length of 35 inches and a weight of about 150 pounds. Although superficially it looks like any other big turtle, it belongs to a primitive family, the side-necked turtles, which were the dominant freshwater turtles of the Cretaceous and Paleocene but are today restricted to South America, Africa, and Malagasy. Instead of pulling the head straight back into the shell as most turtles do, these turtles retract it by bending the neck laterally, a less efficient protective system, for it leaves more of the neck vulnerable to the teeth and claws of predators. Strictly aquatic, slow, and inoffensive, the Arrau turtle is a vegetarian. From January to June when great areas of the rain forest are inundated, it feeds principally on fruit that falls into the water.[29]

From unlaid egg to 100-pound adult, the Arrau turtle is a most important resource for the riverine Indians of South America. Probably no species except some sea turtles has been so relentlessly exploited by so large a human population. The adult turtles are harpooned, trapped, or caught by hand whenever encountered. The greatest slaughter takes place during the nesting season. As has been mentioned, this species is one of those strongly committed to communal nesting on certain river beaches. In spite of its dangers, the habit has some advantages. It insures that the eggs will be laid where and when the conditions of temperature, moisture, and soil consistency will be optimum for development of the embryo and emergence of the hatchling. And if enough eggs and hatchlings are provided, the natural predators will be glutted before the supply is exhausted, thus allowing some of the young to grow to a size at which life is less precarious.

The numbers of the Arrau turtle were once so great that a fair-sized human population could be supported by their bounty. During September and October, when the turtles migrate to their nesting grounds, the rivers were said to become so crowded that the sound of the turtles' shells bumping together could be heard for a considerable distance, and it was difficult to paddle a canoe through the mass of migrating chelonians. Once on the nesting beaches, each female deposits 80 to 200 eggs. These are relentlessly gathered by the Indian population.

About 1850, the English naturalist and explorer Henry W. Bates watched some 400 people assemble for an egg hunt in response to an announcement on the church door. Carrying kettles and jars, they went to the sandbar and formed a large circle. On signal from a roll of drums, they began to dig. After two days, the mounds of excavated eggs were 4 to 5 feet high, and after the four-day pillage was over, Bates estimated that 48 million eggs had been destroyed. Only a small percentage went directly for food; the greater quantity were crushed and rendered for oil which filled approximately 8000 jars containing 3 gallons each. The Indians use the oil for cooking, as fuel in lamps, and mixed with a type of tar for waterproofing canoes.[30]

In January, when the eggs begin to hatch, the young turtles are gathered for a chutney-type sauce known as *mexira*.[31] In recent years, thousands of these little turtles have found their way to Europe and the United States in the pet trade. As long as the Indian population was comparatively small and living under primitive conditions, the tremendous fecundity of the turtles allowed them to maintain themselves in sizable numbers. Some nesting sites were overlooked; others could not be reached during the critical times of egg deposition and hatching; and it was unprofitable to ship the turtles to distant markets. Recently, however, with the use of aircraft and power boats to locate and reach nesting sites and the shipment of thousands of adult turtles by rail and highway to the larger cities, the Arrau turtles are being seriously depleted. Attempts are being made to save them and still allow them to retain their important role in the economy of the local Indians. One of the Arrau turtles' staunchest champions is Janis Roze, a Latvian zoologist who emigrated to Venezuela after World War II. He has produced an interesting and perceptive film on the breeding migration of these turtles and incidentally on the thoughts and emotions of a zoologist in the tropics. Currently the Arrau turtle is on the international list of endangered species, and its young can no longer be legally brought into the United States. Laws protecting the species in Brazil are on the books, but there is little effort to enforce them.[32]

Another big tropical American freshwater turtle deserves

mention for its size, a respectable shell length of about 2 feet and a weight of 20 to 30 pounds, and also for being the sole survivor of a family of primitive turtles: this is the Mexican river turtle, which has survived from the Cretaceous with very little change. There is nothing particularly striking about its appearance; its unique and primitive features are apparent only to students of turtle anatomy. It is a highly aquatic species with large webbed feet and neck muscles so weak it has trouble holding up its head when out of water. The large nostrils at the tip of its snout are somewhat similar to those of the softshells, and like them it apparently can carry out pharyngeal respiration. It does not bask in the sun but often floats at the surface. It feeds mostly upon aquatic vegetation but in captivity will also take shrimp and fish.[33]

The range of this turtle is confined to the Caribbean lowland of Middle America from central Veracruz in Mexico to Honduras. Within this region it is an important food turtle and can often be purchased in the markets in Guatemala City. One of its common Spanish names, *tortuga blanca*, refers to the white color of its meat, which is said to taste like breast of chicken. It is also often known as *jicotea*, a name applied to a number of different freshwater turtles in the Caribbean region and, corrupted to "hickety," sometimes heard along the Carolina coasts. While the Mexican river turtle is apparently still locally plentiful in parts of its range, such as Lago de Izabal in Guatemala, its restricted distribution and the fact that it is rather easy to catch make it vulnerable to exploitation. It should be given some legal protection before its numbers decrease to the point where its survival is endangered.

8. Giant Land Tortoises

No turtles possess the chelonian virtues of strength, placidity, persistence, and longevity to a greater degree than do the land tortoises. Unluckily for them, they are also highly edible and quite easy to catch. Although the land-tortoise family has some peculiar small representatives, such as the pancake tortoises and hinge-backed tortoises of Africa, the giant species all have much in common. Their shells are prominently arched, and the horny outer plates show concentric growth rings. Their heads are proportionally small and their necks long. The hind legs and feet are not at all turtlelike. The late Karl P. Schmidt had the hind foot of a giant Galapagos tortoise in his office for many years and got much amusement by asking fellow zoologists from what animal it was taken. Nearly everyone guessed that it was from a small elephant, hippopotamus, or rhinoceros; only a few recognized it as coming from a turtle. The forelimbs are heavy and slightly flattened; their outer surface is covered with large, heavy scales. Although these turtles cannot close their shells as some small kinds can, the big, armored forelimbs are used to protect the retracted head. Many of the small and medium-sized land tortoises have handsome shell patterns, but nearly all the big species run to drab, uniform blacks and browns.

The land tortoises have a wide distribution in the warmer parts of the globe, although they never reached Australia and are poorly represented today in Europe and North America. The present headquarters of the group is Africa. In America north of Panama, it is represented only by the four species of gopher tortoises. Three of these, which occur in the southern United States, have been well known for years. The fourth and largest, which almost merits inclusion among the giant reptiles, was not discovered until 1958. It inhabits a small area in the Bolson Basin of north central Mexico. The local people, of course, have long known of it. While they eat it occasionally, they also seem to recognize its rarity and conserve it by not taking the eggs or young and by dumping fresh alfalfa and other green plants near the tortoise burrows.[1] A major threat to survival of the species seems to be the recent demand of zoos and private collectors for specimens.

Although these turtles are the poorest swimmers of any of their kind, they have been spectacularly successful in colonizing certain island groups. The most famous of these island populations is that inhabiting the Galapagos Islands which lie about 600 miles off the coast of Ecuador. Even more remarkable in terms of zoogeography were the giant tortoises of the islands of the western Indian Ocean which are all separated from one another and from the nearest large land masses by distances of 500 to 1250 miles. To make such voyages by purposeful swimming would be totally beyond the tortoises' ability. It is barely possible that they might have drifted to the islands, helped by favorable winds and currents. Some land tortoises have buoyant shells; they are all tough animals, and a single egg-laden female could found a colony. When Captain David Porter of the U. S. frigate *Essex* engaged British whalers off the Galapagos in 1813, the British seamen, clearing their ships for action, tossed overboard the tortoises they had collected for food. A few days later, Porter's crew recovered about fifty of the reptiles floating about in apparently good condition.[2]

The most likely hypothesis, however, at least for the Indian Ocean species, is that the island inhabited by the tortoises are the last remnants of a much larger land mass or archipelago that has now largely vanished beneath the sea. The tortoises would probably have had to cross water gaps, but these would have been less extensive than those existing today. The tortoises that originated the island colonies need not have been giant species, although a big tortoise is likely to be more seaworthy and better able to endure the privations and battering of an inter-island float trip than a small species. If these vegetarian tortoises reached an island with plant life and without mammalian competitors or predators, they would quickly become the dominant land vertebrates. A tendency for large size, which seems to be latent in many land-tortoise stocks, would be favored because large animals would be adapted to low-level browsing as well as grazing. Moreover, large males tend to oust smaller males in competition for females, and large females are more fecund egg-producers.

These big tortoises, like many specialized island animals and plants, are highly vulnerable to changes in the physical environment, as well as to the impact of introduced competitors and predators. Several giant island tortoises, such as Gadow's tortoise on Mauritius and the Barrington Island tortoise of the Galapagos, became extinct for unknown causes before the advent of Western man on their islands. So did the most spectacular species of all, the horned tortoise that lived until a few thousand years ago on Lord Howe Island off the eastern coast of Australia. This grotesque monster had a head nearly 2 feet wide and a pair of stout, laterally projecting horns. No complete shell has been found, but it must have been bigger than any land turtle alive today. It was not related to the other giant island tortoises but was the sole survivor of a very primitive turtle family.[3] It is barely possible that some early human sea rover may have seen a horned tortoise eye to eye and that the beast may still linger enshrined in the myths and racial memories of some Australian or Melanesian tribe.

The greatest disaster for the giant island tortoises has been contact with Western civilization. Their islands were discovered during the great voyages of exploration in the sixteenth and seventeenth centuries. To European mariners the tortoises, fat, slow, and gentle, must have seemed the gift of a benign providence to supplement their unspeakably vile shipboard rations. As trading, whaling, piracy, commerce, and warfare increased sea traffic in the eighteenth and nineteenth centuries, the stolid, lumbering giants of the islands almost vanished. At least eleven island races of large tortoises have been exterminated in the last 250 years; most of the others are so reduced in numbers that their survival is precarious.[4] Most of this decline has been due to unrestricted killing of the tortoises for food, but another important factor was the introduction onto the islands of rats which eat the eggs and young tortoises, goats which compete with the adult reptiles for food, and pigs which do both.

There is surprisingly little data on the maximum size reached by the large island tortoises. Early-nineteenth-century visitors to the Galapagos, such as Captain Porter of the U. S. Navy and Captain Benjamin Morrell, a sealer, estimated the weights of the largest tortoises as 400 to 800 pounds. Commander Cookson of the British Navy, who visited the islands in 1875 for the purpose of obtaining information on the tortoises for Dr. Albert Gunther of the British Museum, recorded specimens weighing 241 and 201 pounds.[5] The California Academy of Science expedition that visited the Galapagos in 1905 and 1906 made the most complete scientific collection of these tortoises that will ever be assembled; it consisted of 237 specimens, from nearly every locality where tortoises had been reported. Incidentally, this expedition was afield at the time of the great earthquake and fire that destroyed the Academy and its collections. The material it brought back forms the nucleus of the present Academy's collections. The largest tortoises the expedition obtained had shell lengths (measured in a straight line) of 40 to 41 inches and were taken at Bank's Bay and Tagus

Cove on Albemarle Island and on Indefatigable and James Islands.[6] The collectors did not record weights of their specimens, so it is difficult to correlate their data with those of earlier writers who recorded weights; however, a 41-inch zoo specimen weighed 450 pounds. While it is probable that the very largest Galapagos tortoises may have vanished into the larders of anonymous whalers and buccaneers, some huge specimens with shell lengths up to 51 inches and estimated weights of 550 pounds were seen on Albemarle Island in 1970.[7] All very large specimens for which information is available have been males.

There is even less reliable information of the maximum size reached by the giant tortoises of the islands in the Indian Ocean. The record seems to be held by an Aldabra tortoise with a shell length of 55 inches and weight of 575 pounds that died in the London Zoo in 1898. The only known specimen of the now extinct Farquhar Island tortoise has a shell length of 49.5 inches, and an extinct Malagasy species was about the same size.[8] The big tortoises of Réunion, Rodriguez, and Mauritius seem to have had shell lengths of not much more than 3 feet, but there are almost no complete specimens of these vanished reptiles.

Several land tortoises of respectable size inhabit the mainlands of Africa and South America. The African spurred tortoise, which is found in the arid country from Eritrea to Senegal, is known to reach a shell length of 30 inches and weight of 184 pounds. The leopard tortoise, with a wide range in Africa south of the Sahara, can attain a shell length of 23 inches and a weight of 69 pounds. In the New World, the South American forest tortoise is known to reach a shell length of 27 inches.[9] The relationship of these mainland forms to the island giants does not seem to be particularly close.

In nearly all ways, the giant tortoises behave like their smaller relatives. Charles Darwin was the first naturalist to observe the Galapagos tortoises on their home islands. He saw his first ones on

Chatham Island September 17, 1835, in an area of small volcanic craters,

> . . . which vividly reminded me of those parts of Staffordshire where the great iron-foundries are most numerous. The day was glowing hot, and the scrambling over the rough surface . . . very fatiguing; but I was well repaid by the strange Cyclopean scene. As I was walking along I met two large tortoises, each of which must have weighed at least two hundred pounds: one was eating a piece of cactus, and as I approached, it stared at me and slowly stalked away; the other gave a deep hiss and drew in its head. These huge reptiles, surrounded by the black lava, the leafless shrubs, and large cacti, seemed to my fancy like some antediluvian animals.[10]

Darwin went on to say that the tortoises were more numerous in the higher, damp parts of the islands and that they beat well-chosen tracks to water holes where they drank and wallowed in the mud. He timed their walking speed as 360 yards an hour or 4 miles a day, allowing for a little time to eat on the road. Like most people who encounter a giant turtle, he could not resist the temptation to climb on its back for a ride but found it difficult to keep his balance.

In October Darwin found tortoises laying their eggs, usually burying them in typical chelonian fashion in sandy soil but sometimes placing them in rocky fissures. The eggs were white and spherical; one he measured had a circumference of 7⅜ inches. He also mentioned that the male tortoise utters "a hoarse roar or bellowing . . . during the breeding season when the male and female are together."[11] Male land tortoises great and small are apt to be noisy during periods of sexual activity. We learned this first hand from Zorba, a male Mediterranean tortoise I picked up on a dusty road in Thrace in 1965. In the spring of 1970, we collected a comely female tortoise in the hills between Tel Aviv and Jerusalem and brought her back to Indianapolis. Zorba promptly began to pursue her and attempted to mount her, at the same time emitting a mewing sound that could be heard throughout the house. Since we presume male tortoises of this group are equally vocal in the wild and have a spring breeding season, the Biblical reference to the voice of the turtle (Song of Solomon 2:12) rests on perfectly good zoological grounds. Zorba, incidentally, has yet to be successful in his suit—at least no eggs have been produced—and we find that he and Raquel, his consort, belong to different although superficially similar species. He is nothing if not persistent, however, and he may yet transcend whatever isolating mechanisms keep the species from hybridizing in their homeland.

The giant tortoises deposit comparatively small numbers of eggs—usually no more than twenty, although a female may lay twice in a season. The nesting pattern seems to have developed in a climate more humid than that the tortoises now inhabit, for even before the coming of man and his commensal animals many young tortoises died because they were unable to dig out of the hard soil over the nest. This may explain why tortoises disappeared from some islands without human intervention.

Growth of the young tortoises is rapid. Aldabra tortoises are reported to reach shell lengths of 18 to 22 inches in 4 years and full growth in 25 years.[12] Galapagos tortoises are said to double their weight annually, and one individual increased in weight from 29 to 350 pounds in 7 years. Two specimens in Chicago's Brookfield Zoo weighed 30 pounds when purchased in 1929 and had increased to nearly 400 pounds in 1955.[13] The longevity record for a giant tortoise, and indeed for any vertebrate, is held by an example of the now extinct Seychelles tortoise brought from those islands to Mauritius by the French explorer Marion de Fresne in 1766. It lived on the island in the artillery barracks at Port Louis until 1918 when it fell through a gun emplacement and was killed. At that time it had been in captivity 152 years. Since it was adult when captured, its true age must have been between 170 and 180 years and possibly more. A giant tortoise, probably from Aldabra, lived on Saint Helena for approximately 120 years, and another individual brought to Saint Helena in 1882 was reported to be still alive in 1967.[14] Life spans of over a century are not restricted to the giant tortoises. A Mediterranean tortoise placed in the garden of Lambeth Palace in London in 1633 by Archbishop William Laud lived, according to one account, until 1730 and, according to another, until 1753. The Peterborough tortoise, another example of this species transplanted to England, is often reported to have lived 220 years, but 92 years seems to be the maximum that can be authenticated for its life in captivity.[15] Tui Malila, a Malagasy starred tortoise allegedly presented to the Queen of Tonga by the English hydrographer and explorer Captain James Cook in 1773 or 1777, survived until May 19, 1966, and thus would have been at least 200 years old.[16] Unfortunately, we can find no account of this incident in Captain Cook's report of his voyages, nor did he touch Malagasy en route to Tonga. This casts considerable doubt on Tui Malila's origins as well as its age. Ages of 120 to 130 years have been reliably established for eastern box turtles living free in New England, on the basis of dates carved in their shells.[17]

The giant tortoises seem to be largely vegetarian. Those of the Galapagos feed mostly on cactus and grass, although those at Tagus Cove were found feeding heavily on the applelike fruit of *Hippomane mancinella*.[18] This is the infamous manchineel tree of the Caribbean region. Contact with its sap or with smoke of its wood when burned causes severe inflammation of the skin and eyes. Its fruit is deadly to most animals, but the tortoises seem to be immune to the poison. Darwin saw them eating the leaves of trees, a berry called guayvita, and a lichen.[19] Some of the Galapagos races, such as that on Duncan Island, have saddle-shaped shells with a high, wide front, and very long necks. This body type is associated with browsing, while the races with the more hemispherical shells, such as the one on Indefatigable Island, feed on vegetation closer to the ground. The Aldabra tortoise seems to be mostly a grazer, although it will eat a wide variety of plant material. Young in captivity are fond of canned dog food and grow more rapidly than they do on a wholly vegetarian diet.[20] The tortoises are more efficient in converting coarse vegetation into highly palatable meat and fat than are mammals of the same size, and this, coupled with their placid and inoffensive nature, has been their undoing.

The old pirate William Dampier wrote of the Galapagos tortoises in 1697: "They are extraordinarily large and fat, and so sweet that no pullet eats more pleasantly."[21] All seafarers who came after him were extravagant in their praise of tortoise meat and oil, perhaps because it was such a contrast to the brine-bitter, mahogany-hard preserved meats that were standard provisions on sailing ships of the time. Differences in flavor of meat from tortoises living on different islands were also noted, those on Charles and Hood Islands being particularly tasty and those of James Island less so. Darwin lived on tortoise meat during part of his stay on the islands; his enthusiasm for it was restrained. The tortoises of the Indian Ocean islands evidently were equally good eating, for thousands were taken from Mauritius, Réunion, Rodriguez, and the Seychelles in the eighteenth century. By about 1800 only the Aldabra population remained. It has survived because Aldabra, although small, is a singularly rough island which lies well off the main shipping routes and was never permanently settled. The tortoise colony has received some degree of governmental protection since 1874, and only recently an attempt to establish an air base on Aldabra was defeated by conservation interests. The tortoises have also been introduced into some of the Seychelles islands where they seem to be doing well in a semi-feral state.[22]

Madge and I visited Mahé in the Seychelles in the autumn of

1962. As we sailed into the harbor, where a British frigate rode at anchor, time momentarily rolled back for me to 1944, and I watched nervously for gun flashes and torpedo planes. However, the sailors on the customs launch might have stepped from a nineteenth-century British print, and the whole tempo of life was more suited to the times of Nelson than of Nimitz. Port Victoria dozed at the base of forested granitic hills, and the town clock struck every hour twice—the first time to wake you up.

The Seychelles must have once been part of a larger land mass, because they have some animals, such as the odd wormlike amphibians known as caecilians, that could never have crossed an appreciable water barrier but that also occur in southern India and Africa. The islands must also have been isolated for a long time, for at the time of their discovery their land plants and animals were highly distinctive. The former included the coco-de-mer, a palmlike tree whose seed is the largest in the plant kingdom. The lizards, snakes, frogs, and caecilians were nearly all unique, although more or less African in their ancestry. So were the giant tortoises that originally occurred on all the larger islands. Unfortunately, this island biota, like so many others, has been almost destroyed by introductions from the continents. We found bits of the original forest on the less accessible parts of Mount Misré, but most of it had been sacrificed for the cultivation of coconuts, cinnamon, vanilla, and patchouli. Some of the native reptiles, such as the little velvet-green day geckos, were common, but others, such as the burrowing skinks and the chameleon, are now rare. We did not see any of the distinctive Seychelles frogs, yet the little introduced Mascarine frog was abundant. At several places on the island, we saw giant tortoises penned up like pigs and being fattened on bananas and table scraps for future dinners, while the souvenir shops in Port Victoria were full of crudely stuffed young and half-grown specimens. However, on the way up Mount Misré we saw a huge tortoise amble placidly across the road and disappear into the underbrush. Doubtless it was no more a wild animal than the bullocks that also roamed the countryside, but for a moment we were carried back in time to the days when the tortoises were the largest land animals on the islands. Incidentally, the tortoise importations from Aldabra to Mahé may have begun before the native wild stock was extinct, so the present-day animals may retain some of the genes of the vanished Seychelles race.

The fourteen or fifteen races of the Galapagos tortoise have been observed rather closely since the California Academy expedition, although efforts to protect them have been only partially effective. That expedition considered four of the five races on Albemarle Island to be abundant to fairly numerous. Albemarle is the largest island in the group and was apparently formed from the fusion of several smaller islands through volcanic upheavals. This explains the presence of several distinct kinds of tortoise. The only other islands where the expedition considered the tortoises fairly abundant were Duncan and Indefatigable. The populations on the remaining islands were rated as rare to extinct.[23]

The most recent evaluation of the Galapagos tortoises is that of Peter C. H. Pritchard, a young British herpetologist, who is an expert on turtles and who visited the islands in 1970. His report is surprisingly encouraging. At Cowley Mountain, or Volcan Alcedo, on Albemarle, where the California party reported tortoises as rare, he saw about 400 tortoises in one morning, some of them the largest ever reliably reported from the Galapagos. They had made an impressive complex of trails within the crater. Copulating pairs and young of various sizes indicated a thriving population. Tortoises were seen on the other four major volcanoes of Albemarle, although on the southern two they were scarce. Tortoises were also quite plentiful on Indefatigable Island and were holding their own on Chatham and James islands. The distinctive Duncan Island tortoise is making a comeback against almost incredible odds. For about 70 years, practically every hatchling tortoise on this small island had been eaten by rats, and many adults captured or killed. In the last five years, nests have been located and dug up shortly before the eggs hatched. The young are raised at the Darwin Experimental Station until they are too large to be harmed by rats and are then released on their home island. What is more remarkable, the tough old adult reptiles seem to be outlasting their rodent enemies. The rats are dying out, and a few wild-hatched young tortoises are beginning to appear. The long tortoise life span does have advantages! The Hood Island population is down to seven known animals. Five have been transferred to the Darwin Experimental Station where they are breeding, and the race may yet be saved.[24] The California Academy expedition seems to have gotten the last specimens from Jervis, Abingdon, and Narborough islands to reach scientific collections, but there is a chance that a few remain on the latter two islands. The tortoises of Charles and Barrington islands have been extinct for a long time.

Conservation of the unique Galapagos fauna is now in the hands of a multinational group at the Darwin Station, and salaries of two rangers are being paid by Lindblad Travel, an organization that specializes in tours to unusual and biologically interesting places. There is now reason for some cautious optimism about the future of the surviving races of the Galapagos tortoises.

9. Legendary Turtles

The Mediterranean Basin

Turtles were very important to me in the winter of 1925, probably because I was a fragile little girl and the object of kindly concern by one Mrs. Ahnspaugh, who brought pots of turtle soup to ease my growing pains and "put some meat on them bones." Mr. Ahnspaugh was the town barber, but barbering was strictly a Saturday affair in the farming community in southern Indiana where we lived. His true vocation was turtle hunting.

As for me, whooping cough, scarlet fever, measles, and pneumonia each followed hard upon the other, and Mrs. Ahnspaugh's turtle soup was probably fully as helpful as the nostrums in our family doctor's black bag, there being no sulfa drugs, antibiotics, or protective serums. I spent long afternoons convalescing at the living-room window where I watched for Mr. Ahnspaugh to come tromping down the road, his boots heavy with clay and the poke (gunny sack) slung over his shoulder bulging with softshells and snappers.

The best times were when I could cross the grass-paved alley between our houses and watch him unload his catch. Sometimes he let me choose which turtle he should kill and clean for supper. I always chose the biggest because I knew that Mrs. Ahnspaugh would want the shell. When it was well dried, she would carefully sand it, enamel it a brilliant emerald, and paint huge red and pink cabbage roses on it. Then she would hang it in the parlour.

Several decades later when I read Pliny's *Natural History* I found that Mrs. Ahnspaugh's friendly pot of turtle broth had classical antecedents, although to be wholly effective the turtle should be collected at the full of the moon and decapitated with a brass knife. Thus caught and killed, the turtle's meat and broth might also cure the king's evil (scrofula) and drive away fits. And the medicine would be even more effective if the turtle were seethed in wine with beef gall and snake-sloughs (shed skins). Other ancient prescriptions included the use of turtle gall, "mix't with Atticke honey," to treat a scorpion sting, and turtle shell, ground and combined with fat, wine, and oil, to cure ulcers of the feet.[1]

Roman women sometimes dyed their hair yellow with sea-turtle gall, rubbed their teeth with turtle blood to whiten them and prevent toothache, and scraped turtle shell into the drinks of over-ardent swains to cool their lust.[2] Roman mothers dropped a mixture of turtle blood and breast milk into aching stopped-up ears, "an excellent medicine to scour them," and treated eye infections with turtle gall—a broadly useful preparation, since it could also be rubbed on swollen tonsils to ease their pain and used by men for "inflammation of the genitors."[3] Turtle blood, taken in a new earthen vessel never used before, was considered a specific for shingles, erysipelas, and warts. A few drops in a patient's nostrils would sooth an epileptic fit.

Pliny recommended sea-turtle blood for baldness:

> The blood of the sea tortoise serveth to recover hair in places naked and bare, by occasion of the deasise called Alopecia. It riddeth away likewise scales and dandruff, yea, and healeth all the scalds of the head, but the same must dry upon the head and be washed off at leisure, little by little.[4]

Before being too disdainful of the many queer nostrums in the Roman and Egyptian pharmacopeias, it may be well to reflect that modern medicine has derived some of its most potent therapeutic substances from the testicles of bulls, the pancreas of hogs, the kidneys of baboons, the urine of pregnant mares, and the mold on a cantaloupe in a Peoria supermarket.

South across the Mediterranean in Egypt, the turtle was not valued as medicine nor did it enjoy the semidivine status accorded the other reptiles—snakes and crocodiles. It was classed with fishes, creatures of watery darkness, and considered one of the enemies of the sun god, Ra. On a tomb from the Nineteenth Dynasty (1320–1200 B.C.), the deceased is depicted as harpooning a Nile softshell, accompanied by the hieroglyphic, "May Ra live and may the turtle die."[5] Reliefs in temples of Philae, Dendera, and Edfu show this turtle being ritually harpooned by the pharaoh. A Middle Kingdom coffin text of about 2000 B.C. contains a spell protecting the dead against the distressing possibility of having to eat excrement in the next world: "If you tell me to eat this, then Ra will eat turtle"—placing the turtle firmly in an unsavory category.[6]

Sea turtles were highly regarded in the Mediterranean basin in preclassical through classical times. In the eighth century B.C. the citizens of Aegina struck the first coins of European Greece in honor of Aphrodite, goddess of sexual love and beauty. On one side the coins showed a sea turtle, sacred to the goddess and also the badge of the city. These staters, known as "tortoises," were the principal medium of exchange throughout the Peloponnesian world long before Athenians took to the sea. Even after Athens began to distribute her own coinage, showing owls sacred to Athena, the tortoise coins were preferred because of their greater weight, being known as the "heavy drachms." About 500 B.C., for reasons as yet unknown, Aegina stopped issuing coins showing a sea turtle and struck coins depicting a land tortoise, an animal sacred to Hermes, messenger of the gods. Perhaps this displeased Aphrodite; at any rate, in 453 Aegina became a vassal to Athens, and Aeginian coins disappeared from international markets.[7]

India and the Far East

Across Asia and into the Far East, turtles and tortoises have been valued less for their medicinal properties and more for their magic powers and religious significance.

Turtles seem designed by nature to bear burdens, and it is in this capacity that man has assigned them significant roles in his many myths of creation. In India, China, Japan, and North America huge supernatural turtles support mountains, islands, continents, and, in the case of Hindu cosmology, the entire universe.

Judeo-Christian time schedules for creation seem cramped and meager when compared with India's extravagant vision, in which the universe dissolves and is re-created every 4,320,000,000 years.[8] Vishnu, second god of the Hindu triumvirate, transforms himself into a huge turtle. Solidly anchored in space by the force of his omnipotent will, he bears on his back the cosmic churn which gods and demons will use to re-create the world. Once the task is completed, Vishnu's turtle avatar remains the keystone of creation, holding on its back a huge elephant, on whose back, in turn, the universe rests.[9]

Giant mud turtles, which may be any of the Indian softshells, have played an essential role in Vedic sacrifices in India for more than 3000 years. In preparation for the rite, a hollow altar is prepared, generally built of bricks. It is constructed around a depression in which a turtle with a supply of food is placed before the altar is capped. After the celebration, the altar is opened, and if the turtle is still alive, the sacrifice is regarded as auspicious. If the animal is dead, the gods have ignored the offering.[10]

Tortoises of various species have roamed freely in our house for years, and in Pakistan we were pleased to learn that this custom not only insures household prosperity and long life but has royal precedents. In the Vrhat Samhita of Varaha Mihir, a Sanskrit encyclopedia of the sixth century A.D., kings and princes are advised to rear up tortoises and turtles in their palaces.

> The shape should be like that of a water jar, with a beautiful bridge at its back; or it may be of the rosy colour like that of the morning sun, with spots like black mustard seeds. If such a tortoise is kept in the house it increases the greatness of the king. . . . The tortoise which has a body black like khol [eye-paint], or like the bee variegated with spots, which has no defective limbs, and whose head is like that of a serpent and the throat thick, increases the prosperity of the empire.[11]

In ancient China, Kwei, the divine tortoise embodied in the constellation Ursa Major, was Lord of the Northern Quadrangle. He was called the Black Warrior[12] and was the longest-lived and kindliest advisor to man of all the supernatural creatures who presided over creation and the fate of China. For 18,000 years Kwei supervised the formation of the earth, the placing of the sun, moon, and stars, and the expansion of heaven.[13] After all was completed, Kwei took thought and by thinking conceived. Shortly thereafter he produced the great millennial turtles of China and Japan, who seek truth, assist mankind, and support the world's burdens. One holds up the subcontinent of India; another bears Horai, the island of paradise, on its carapace.[14] When one of creation's four pillars collapsed, yet another great turtle placed itself under the sagging corner to steady the tipsy universe.[15] One legendary tortoise assisted China's first emperor, Shih Huang Ti (221–207 B.C.), by offering itself as a foundation for a great wall which the emperor planned to build. The wall is still known as the Kwei Ching, or tortoise wall.[16]

One millennial turtle heard the "Blessed One" and was converted to Buddhism. It roams the sea at depths where it needs no eyes since light never penetrates. However, it does have one eye, which is centered in its plastron like a navel. Every 3000 years it ascends and surfaces in order to see the sun. First it blindly searches for a floating plank large enough to bear its weight. This plank must have an open knothole through it. The turtle hauls out onto the plank, places its eye over the hole, and waits for a storm. When a storm finally strikes, heavy seas turn the plank with the turtle upside down, so that when the skies clear and the sun breaks through, the turtle sees the sun through the knothole in the plank. Satisfied, it sinks back into the depths to wait another 3000 years. Such, say the yellow-robed priests, are the difficulties intrinsic in seeing the light.[17]

In Japan, the tortoise is one of the familiars of Benzai-ten, goddess of conjugal love, learning, and eloquence, a Japanese equivalent of Aphrodite and her sea turtle.[18]

Among the more poignant stories concerning supernatural turtles in the Far East is that of Urashima, the Japanese Rip Van Winkle.

In the twenty-first year of the reign of Yur'yaki-Tenno (487 A.D.), Urashima, a fisher boy, went to sea alone in his boat to try his luck. For three days and nights he fished without success. At length he caught a turtle of five colors. Wondering but grateful, he pulled it into the boat and fell asleep. While he slept, the great turtle was transformed into a woman beautiful beyond description. When Urashima woke up he was amazed. Smiling, the lady said, "I deemed you a man of parts alone on the sea, lacking anyone with whom to converse, so I came here by wind and cloud."

Presently their boat drifted to a broad island covered with jewels. As Urashima and the woman approached the gates of the city, a horde of charming children rushed out to welcome them, crying, "This is a husband for Princess Turtle!"

Despite Urashima's reservations about the woman, he wed her and was content for three years before he began to mourn for his parents and native village. Sadly the princess confessed that she was indeed the great turtle which he had caught. She gave him a jewel box, promising him that he might return to her if he would just not open the box.

When Urashima returned to his old home he could find no trace of family or friends, and not many familiar places. Of course he opened the box and then realized, as his body withered into dust, that he had been away not three but three hundred years.[19]

Urashima was the ancestor of the Kusaka clan. The English scholar Basil Hall Chamberlain, in his *Classical Poetry of the Japanese* (1880) says: "Urashima's tomb, with his fishing line, the casket given him by the maiden, and two stones said to be precious, are still shown at one of the temples in Kanagawa, near Yokohama."[20]

When the German traveler Engelbert Kaempfer visited Japan in the late 1600s, he reported:

> Of all the footed animal produce of the water, the Ki, or Came, Tortoises, are most esteemed by the Japanese, being looked upon as peculiar emblems of happiness by reason of the long life which is ascribed to them. That kind particularly which hath a broad tail much like a large round beard and which in their learned language is called Mooke . . . is frequently to be seen among other emblematical figures wherewith they adorn the walls of their temples, the sides of their altars, and the apartments of the Emperor and the Princes of the Empire.[21]

Today in the many holy temple pools where tortoises are well fed by pilgrims and devotees, they achieve great age. On the carapace of an old specimen algae and parasitic plant growths may form a kind of beard. Such a turtle enjoys general interest and goodwill.

In old Japan it was believed that turtles procreated only by thought, but according to Chinese popular belief all tortoises and turtles are female and, in order to reproduce, mate with snakes. In both cultures, "progeny of the tortoise" is a euphemism for bastard. In modern China it is still a blatant insult to call a man a "turtle egg," and a subtle one to tell him to "roll away."[22] This ambiguity toward the anciently venerated tortoise did not exist before Confucius. Legend has it that the great philosopher taught that the tortoise carried a record of the Eight Virtues on its back because otherwise it could not remember them. Thus he stigmatized the animal for its irresponsibility and impugned its morality.[23]

North America

When Asian man crossed the Bering land bridge and began to filter south into the American continents, he brought the millennial turtle with him. The Walam Olum, tribal chronicle of the Delaware Indians, describes the earth as an island tied to and supported by a huge turtle. The saga, written down in 1833 by an early American naturalist, Constantine Rafinesque, is told in pictographs said to be markedly similar to archaic Chinese symbols. In total length it runs to 183 verses, relating the tribal story of Creation and the Lenni-Lenape migration from Asia to Alaska and south to east across North America. In verses 14 and 15 of book two, the turtle-borne island offers the people safety during a great deluge[24] which may represent the floods released by the last melting ice sheet, 10,000 to 12,000 years ago.

Kwei, Lord of the Northern Quadrangle in China, has a counterpart in the American Southwest. The Navajo Indians in New Mexico and Arizona venerate Sistyel, the tortoise, as the guardian of the northern door of creation's first house. The parallelism goes even further; Sistyel is associated with the blue heron just as Kwei is often paired with the crane, a bird highly respected in China for its interest in human affairs and its fabulous longevity.[25] And the Chinese Blue Dragon of the East is echoed in Navajo mythology as the water monster, Tieholtsodi, who represents the east and lives in a house of clouds guarded by Icni, the thunder.[26]

Wyandot Indians claimed to be the first tribe engendered from the earth. According to their traditions, Atahensic, the queen of heaven, was discovered in a love affair with one of the six gods who rule the sky. Her enraged husband, Atahocan, hurled her from the walls of heaven. Below, the great tortoise raised his carapace to receive her. Soon after landing she gave birth to twin sons, Inigoria (the Good Mind), and Inigohatea (the Evil Mind). The tortoise grew to immense size and became the home of all men.[27] A Cherokee legend adds that after giving birth to the twin boys the goddess died. Her head became the sun and her body the moon.[28]

The Turtle Clan of the Iroquois claimed descent from a giant turtle who, burdened by the weight of his shell in walking, contrived to throw it off and thereafter gradually developed into a man.[29]

Many of the Algonquin people formed their camps in the "shell of the big turtle," beginning at the right foreleg and continuing on around the shell. Families of the turtle totem camped at the turtle's head.[30]

According to the shamans of the Iroquois nations, before the earth was made, man lived in the sky-beyond. There was no night, light being provided by the blossoms of a dogwood tree which grew beside the chief's lodge.

Now the chief was a churlish man, hypertensive and obstinately sulky. His wife tried not to provoke him but he was often angry with her. One day, in a savage humor, he seized the dogwood tree and pulled it up, breaking a hole into the sky-beneath. Disconcerted by what he had done, he lay down beside the hole and pretended to be sick. The dogwood blossoms began to droop, and the light faded. When his wife came to comfort him, the chief induced her to sit beside him and dangle her feet into the hole. After she was comfortably seated, he seized her by the neck and pushed her into the abyss. She fell like a meteor through space.

The White Fire Dragon of the Storm handed her an ear of corn, a mortar, and a pestle as she passed, and the totemic animals of the sky-beneath hastily met to debate who should catch her and how this might be done. Finally, Turtle was selected, and he spread his broad carapace under her approaching body. She landed softly, her fall cushioned by the wings of Wild Duck. Exhausted, and alternately weeping and sleeping, she rested on Turtle's shell. Otter, Beaver, and Muskrat dived into the cosmic sea to find the bits of magic earth that had fallen from the dogwood tree when it was uprooted. They brought handsful of it to Turtle's shell where Loon and Bittern spread it about and planted willows.

When the woman awoke, a small world was ready to cheer her and she was pleased. She walked about, tracing rivers and planting the corn which the Fire Dragon had given her, and she blessed the earth so that it increased rapidly in size and fertility.[31]

There remained one serious problem. Earth was dim, and the corn could not sprout without light and heat. Turtle called a second summit meeting of the animals and they discussed how this difficulty might be solved. Small Turtle offered to climb the hazardous trail up to the sky-beyond if the other animals would help her with their magic. All agreed, and by means of their magic they created a black cloud full of clashing rocks. Small Turtle climbed into the cloud. As the stones cracked together, they generated lightning which Small Turtle gathered and formed into a large ball. When the ball was almost too large for her to control, she threw it up into the sky. Then she collected more lightning and formed a smaller ball, which she threw after the first. The large ball became the sun and the smaller the moon. Turtle instructed Groundhog and Badger to burrow holes in the sides of the sky-beneath so that the sun and moon could go into one and emerge from the other. Thus day and night were established.[32]

10. The Great Lizards

Monitors

The basic evolutionary strategy of the lizards has been the same as that of one of the more successful automobile manufacturers—"think small." During an existence of almost 200 million years, the group has produced comparatively few giant species, although today it is the largest group of reptiles in terms of numbers of species and almost certainly in numbers of individuals as well. Yet among the approximately 3000 species of living lizards there are no more than a half-dozen that top the 6-foot mark in length and probably not more than a hundred that exceed 3 feet. Even among these larger species, many are lightly built, with long, whiplike tails, and some are completely legless and snakelike.

The fossils of the earliest lizards are difficult to distinguish from those of other small reptiles. Such ancestral forms as *Prolacerta* and *Pricea* are found in the welter of reptilian bones discovered in the early Triassic of South Africa, which at that time was fused with Antarctica as a part of old Gondwanaland. By the late Triassic the lizards had spread to the northern continent of Laurasia, and by the Jurassic the group was well established.[1] Most of its diversification seems to have taken place since the end of the dinosaur era, although the fossil record of the smaller types is far from adequate.

Nearly all the giant lizards of today belong to one family, the varanids, or monitors. The name monitor for these lizards seems the result of some linguistic blundering going back to a German writer who incorrectly translated *ouaral* or *ouran,* an Arabic name for one of the common African species, as *waran* or *Warneidechse*—that is, "warning-lizard." This, in turn, became "monitor" in English. The monitors are an ancient group of lizards with a fossil record going back to the late Cretaceous, about 80 million years ago. Their relationship to the extinct marine mosasaurs has already been mentioned. The biggest member of the group was *Megalania*, which reached a length of 20 feet or more and survived in Australia into Pleistocene times, less than a million years ago.[2] Its extinction seems to have coincided with that of some of the giant marsupial mammals and was probably related to the increasing aridity of the island continent in the Pleistocene.

Students of reptile evolution generally agree that the monitors arose from stock similar to that which produced the snakes. One of the most striking similarities between monitors and snakes is in the form of the tongue. In both it is long, highly protrusile, and deeply forked. Its function seems to be the same—namely, to pick up particles from the environment and carry them to a specialized chemosensor in the roof of the mouth known as Jacobson's organ. On the other hand, elongation of the body and reduction of the limbs are far less evident in monitors than in several other lizard families, such as the skinks and anguids. While many kinds of small lizards have tails that break readily if seized by an enemy and just as readily regenerate, this is not the case with monitors. For most of them, the tail is a propeller when in water and a weapon on land.

While there is some specialization of life styles among monitors, especially in regions such as Australia that have a variety of species, most monitors are versatile reptiles that can climb, swim, or run as the occasion demands. Several species are primarily arboreal. I saw my first wild monitor during World War II, near Hollandia in what was then Dutch New Guinea. I was walking through fairly dense underbrush when I heard a slight rustle, and an elongated reptilian head with flickering tongue popped out at about a level with my face. I first thought it was a python but soon realized my mistake. I probably could have caught the lizard, but my shipmates would not have welcomed it alive, and it would have taken most of my supply of formalin to preserve it. I suspect it was the green tree monitor which is common in that region and unique in having a prehensile tail which aids it in climbing.

Several species of monitors swim well, propelling themselves with the tail in the manner of crocodilians. One species has been reported to be able to remain submerged an hour. The Indian monitor, a moderate-sized species that reaches a length of about 4 feet, is common along the canals and sluggish creeks of the Indus delta and often dives into the water to escape. I saw one in a large well, where the local people told me it had been living at least a year. The two young sons of one of our American friends in Karachi were quite keen about reptiles and kept as a pet a monitor named Amos. The boys' mother was an exceptionally tenderhearted person, and whenever she heard Amos scratching and thrashing about in his cage, she would feel sorry for him and let him out on the veranda. Getting him back into the cage was another matter altogether; he usually escaped into some inaccessible corner of the house, only to turn up a day or so later in the laundry basket or under one of the beds. Finally, however, he leaped into the toilet and plunged down the drain. The family were decidedly uneasy, for they entertained Pakistani and American officials frequently and had visions of a guest's being disturbed at a critical moment by the sudden emergence of a yard-long lizard. However, Amos was seen no more. Doubtless he made his way down the drain to the open sewer and thence to freedom.

A running monitor has a clumsy, waddling gait, but it is usually fast enough to outrun a man in the sort of terrain in which monitors live. The sand goanna of Australia is said to be incredibly speedy. When cornered, it sometimes rears up on its hind limbs.[3]

Monitors are very strong, and it is difficult to extract even a small one from a hole or crevice. Several times while we were in Asia we heard stories of monitors being used as living grappling hooks by burglars or by soldiers attacking walled villages. The lizard with a rope tied around its body just in front of the hind legs is started up a wall. When it reaches the top, the man holding the rope gives a tug, the lizard digs in its claws and holds on, and the man pulls himself up the rope. We doubt that even the biggest Indian monitor could support a man's weight in this manner, although it might accidentally twist the rope around some projection and thus gain additional purchase. However, the larger and much more strongly built Malayan monitor might be able to do so. To what extent this trick is actually used, we do not know. It has many of the earmarks of a folk tale, yet there is nothing intrinsically impossible about it.

Australia, New Guinea, and the nearby islands of Indonesia are headquarters for the monitors and home of about 75 percent of the living species, which range from giants of 10 feet to little fellows no more than a foot when full grown. In Australia, monitors are called goannas, probably a corruption of iguana. Americans serving in the Southwest Pacific in World War II turned the name back to iguana assuming an entirely unfounded relationship between them and the large lizards of the American tropics.

Biggest of the Australian monitors is the perenty, or giant goanna, an uncommon species but one that is widely distributed in desert areas where it lives mostly in rock crevices. It reaches a length of 8.5 feet but is of slender build.[4]

The other giant Australian monitor is the lace monitor, so called because of its attractive pattern of yellow reticulation on a black background. It also is a slender species, reaching a length of a little over 7 feet. It is a tree lizard; hence its range is complementary to that of the giant goanna. It is the most plentiful and widely distributed of the large Australian monitors.

Largest of living lizards is the Komodo dragon, a giant monitor with a limited range in the Sunda Islands of Indonesia. It is best known from Komodo, a rugged volcanic island about 22 by 12 miles in extent, situated between the large islands of Sumbawa and Flores, but is also known from two smaller islands, Rintja and Padar, and from a small area on Flores.

This huge lizard was not recognized by science until 1912, although for some years previous traders and fishermen visiting the Sundas had brought back tales of a strange monster known as *buaja darat*, or land crocodile. It was said to reach a length of 20 feet and be strong enough to pull down a buffalo. In fact, the lizard does well to reach half this size; the largest specimen on record is one of 10 feet 2 inches acquired by the St. Louis Zoo in 1933.[5] However, it is of extraordinarily massive build when full grown. The record specimen mentioned weighed 365 pounds, and adults between 250 and 300 pounds are regularly found. There is evidently a great deal of variation in weight, for a vigorous 9-foot specimen may weigh only a little over 100 pounds.

As the largest land carnivores, these lizards dominate their island world. The larger adults seem to have home ranges of 1 or 2 square miles, which they know intimately. Here they prepare resting places under thick vegetation, where they sleep and seek shelter from the sun, which during the dry season may drive the temperature in open areas to over 120° F., far above the lethal point for any reptile. They also dig burrows. They are usually active only for short periods in the early morning and late afternoon. Despite their bulk, they walk with the body held well off the ground.[6] Young lizards spend most of their first year in trees, and the half-grown ones climb occasionally, but adults are too heavy for such acrobatics. They visit water-holes to drink and to cool themselves and have been reported to evade pursuit by running into the sea.[7] They sometimes swim through swift tidal currents to islets up to 1,300 feet off shore to feed on the domestic goats pastured there.

The largest lizard of mainland Asia is the Malayan monitor, also frequently called water monitor, or *kabara-goya*. Its range extends from Ceylon and the southern part of India eastward to extreme southern China, the Philippines, and many of the islands of Indonesia. Lengths of more than 9 feet have been claimed for this species in the older literature, but more recent authorities give maximum lengths between 2.07 meters and 2.36 meters (6 feet 9.5 inches and 7 feet 9 inches).[8] It is heavily built, and large specimens weigh 50 to 75 pounds. This is a lizard of tropical lowlands, seldom found far from water and often plentiful along the edges of streams. It may well be the most aquatic of all the monitors and has even been sighted well out at sea. Despite its size, it is by no means uncommon in thickly settled regions. In East Pakistan (now Bangladesh) I saw a big one go swishing away through tall grass as I was walking along the edge of a pond near Chittagong, and one of our friends saw a huge one on the grounds of Dacca Medical College.

The Nile monitor is the third of the triumvirate of giant reptiles that were first associated in European minds with that historic river, although, like the crocodile and the softshelled turtle, it ranges over a great part of Africa. Four to 5 feet is its average length, but there are a few reports of specimens of 6 feet or more. The Nile monitor prefers the vicinity of water and is an excellent swimmer.

Among the smaller lizards, sex differences in color and pattern are nearly as obvious as among birds, with males usually being more vividly and brightly marked. This is not true of monitors, however. Not only are the sexes superficially alike, but there is a confusing similarity in the sexual organs. At the Postgraduate Medical Centre in Karachi, the physiologists sometimes used monitors as experimental animals, for they were readily obtained and lent themselves well to experimentation in cardiac physiology. One day a student came to me with the news that all our monitors were hermaphrodites. He then proceeded to show me a dissected specimen, its body cavity crammed with fully developed eggs, but with a pair of penislike organs at the base of the tail. On closer examination, these did indeed appear to be hemipenes, which are the copulatory organs of snakes and lizards, although they were only about half the size of the organs of an adult male. I confirmed this finding on a couple of other egg-containing females, but we have no idea whether it holds true of other monitor species, or if the organs have any function in the female.

There seems to be no great size difference between the sexes in monitors, although observers of the Komodo dragon have reported that the largest individuals are males. In a small series of Indian monitors we studied, the largest individual was a male, but females averaged a trifle larger.

The Nile monitor and the lace monitor are among the reptiles that lay their eggs in the large mound nests of termites. Quite by coincidence, temperature and humidity within a termite mound are ideal for incubation of many kinds of snake and lizard eggs, and for reasons that are not fully understood the insects do not molest the eggs or hatchlings. How the hatchlings make their way to the outside is also not clear.

Young monitors are among the most attractive of lizards, showing intricate patterns of spots and bands on a skin whose texture resembles fine beadwork. Most species tend to darken and lose their markings as they grow older. An Indian monitor that we collected when it was probably no more than a few weeks old grew to adult size in 30 months. Growth of the giant species may be comparable, although we have found little information on this. An adult male Komodo dragon living under natural conditions grew about 6 inches during a period of a little less than 8 months.[9] The life span of the larger monitors is believed to be about 15 years.

While monitors are without exception carnivorous, their range of animal prey is limited only by size and availability. The stomach of a young adult Indian monitor we shot near Karachi contained a handful of large ground beetles, a grasshopper, a big solfugid or wind scorpion, and an adult Indian garden lizard. Other monitors of this species had eaten squirrels, shrews, nestling birds, snakes, fish, crabs, and locusts. Eggs, be they bird or reptile, seem to be the one food that no monitor can resist. Elsewhere we have mentioned the toll that monitors take of crocodile and sea-turtle eggs; they are equally fond of bird eggs. Chicken farmers in Borneo consider the Malayan monitor their number one enemy; it eats everything from eggs to full-grown fowl. Komodo dragons seek out the unique mound nests of the megapodes, or brush turkeys, and dig up the eggs. Lace monitors make a specialty of robbing bird nests, although the big Australian cockatoos sometimes succeed in driving them away.[10]

Adult Komodo dragons hunt much as the larger cats do. Hidden in underbrush beside a game trail, they wait until a deer or wild pig passes by. With a sudden lunge, they catch the animal by the leg or throat, then throw it to the ground and disembowel it. One lizard killed a 40-pound pig in 8 minutes, then proceeded to eat the viscera and finally the carcass. Fifteen minutes after the animal was killed, only a few scraps remained. Even water buffaloes weighing 1000 pounds or so have fallen prey to these giant lizards. The reptiles seize the buffalo by the hind legs in repeated attacks until their sharklike teeth cut through the tendons; then they eviscerate the crippled animal. If an animal tears free of the lizard's jaws, it may die later from hemorrhage or infection and furnish a meal for other dragons.[11] Scavenging seems to be a common trait among the larger monitors. Komodo dragons come readily to carrion baits, and we have seen Indian monitors feeding along with vultures on the carcasses of bullocks and donkeys. On Manus Island in the Admiralties and on New Guinea during World

War II, I was told that "iguanas"—undoubtedly monitors—fed on the bodies of troops killed in jungle skirmishes. Malayan monitors have been seen feeding on drowned cattle and sometimes steal the bait from fish traps.[12]

One of the traits monitor lizards share with snakes is the swallowing of large prey entire. A 101-pound Komodo dragon ate a 90-pound wild pig, which left its belly grotesquely distended and dragging the ground.[13] We had a desert monitor only about 30 inches long that could swallow almost full-grown house rats.

Although the food preferences of large monitors bring down on them the wrath of chicken farmers, in some other ways they may be an economic asset. In West Africa the Nile monitor feeds heavily upon the giant land snail (*Achatina*)[14] and may be important in controlling this big gastropod, which has been a serious agricultural pest in regions such as Hawaii and south Florida where it has been introduced. In Ceylon the Malayan monitor is protected because it preys on the freshwater crabs whose burrows otherwise undermine the dikes, or bunds, surrounding the rice fields. Monitors also help in controlling rats and other undesirable rodents and some types of large insects.[15]

A characteristic defensive reaction of monitors is lashing at an enemy with the tail. This is often accompanied by inflation of the throat and body and sometimes by hissing. The performance is mostly designed to intimidate, and there is a sharp difference of opinion among herpetologists as to how much harm the tail blow of a large monitor can do. Ditmars said a large Malayan monitor in the Bronx Zoo attacked its keeper and inflicted "as severe a welt as if dealt with a whip."[16] Pope, however, said of the same species, "Experiments . . . convinced me that the tail is unable to inflict appreciable damage on a well-clothed man.[17] Zdenek Vogel, director of the herpetological station at Suchdol, Czechoslovakia, warns that: "Even the tamest monitor sometimes lashes out with the tail quite unexpectedly . . . so the keeper should watch out for his eyes!"[18] I have been slapped several times by the tails of 2.5- to 4-foot desert and Indian monitors without anything worse than a slight sting, but I have no desire to repeat the experience with one of the 6- to 7-foot giants. Curiously enough, tail lashing has not been reported as part of the defensive repertoire of the Komodo dragon, the species that is most likely to be able to do real damage.

When defending themselves at close range, monitors bite and claw, and here there is no disagreement about the big species' being formidable antagonists. Even moderate-sized species are no fun to deal with. We once left three Indian monitors in a bag in our jeep station wagon for an hour or so. During this time, they tore a hole in the bag and escaped. Two were recaptured without trouble, but the third had got behind the dash and thoroughly wedged itself in the maze of braces, wires, and rods. As I was trying to extricate it, it seized my thumb in its jaws. Whenever I tried to pull free, the lizard bit down harder, and the space was too narrow for me to use my other hand to pry the jaws open. Luckily Madge was close by, and I shouted to her to fill a syringe with nembutal solution and inject some into the lizard's belly. Soon the reptile relaxed, and I got both it and my thumb free. I had several tooth punctures but escaped infection, which is reputed to be common after monitor bites.

Dr. Walter Auffenberg, an American herpetologist whose field studies on the Komodo dragon are still in progress, says that these giant lizards occasionally attack man without provocation. He recorded several incidents of natives injured while working in the bush or sleeping on the ground during the day. One person was quickly killed; another died four days later of infection; still others had severe injuries from which they recovered. Members of the Auffenberg expedition were attacked by lizards when in tents and blinds as well as in the open but sustained no serious injuries.[19] The Komodo dragon's aggressiveness may be the result of its being the only large predator native to its environment. No such attacks on man have been reported for the other large monitors.

The Komodo dragon has enjoyed governmental protection almost from the time of its discovery, but other large monitors have not been so fortunate. Skins of the large and medium-sized species in quality of leather are second only to crocodilians among the reptiles. Hundreds of thousands of hides are shipped each year from southeastern Asia and a large number from Africa. Skins of lizards used in physiology experiments at the Postgraduate Medical Centre were one of the fringe benefits of laboratory technicians in that department, who had no trouble selling them. The lizards' fat also found a ready market among the hakeems, or practitioners of traditional Indian medicine. Since the monitor, locally called *gho*, has exceptionally keen vision, oil rendered from its fat is used to treat failing eyesight in the aged. Many residents of the Australian outback believe the oil from the goanna has such marvelous penetrating power that it will seep through a glass bottle.[20] In Ceylon, flesh of the monitor is considered a remedy for persistent vomiting.[21]

The blackfellows of Australia, the Dyaks of Borneo, and the aboriginal tribes of Ceylon and southern India are among the peoples who use the meat and eggs of monitors for food. The presence of the Pacific monitor on scattered small islands in Micronesia seems best explained by assuming that it was carried as food by Polynesian sea rovers and either was deliberately introduced or escaped. American field zoologists, such as Karl P. Schmidt and Robert Inger, who have eaten monitors in the course of their travels have compared the meat to that of an old, tough rooster and considered the eggs inferior to hens' eggs.[22]

Certain tribes of Ceylon were said to make a powerful poison known as *kabaratel* from the blood of the Malayan monitor mixed with crude arsenic and cooked in a human skull.[23]

The Selangor Malays say that when crocodile eggs hatch, the mother tries to eat all the young that run away from the water, but those that escape become monitor lizards.[24] In Egyptian folklore, the monitor sucks milk from sheep and goats, thus taking a role usually ascribed to snakes.[25] The people of Sind believe that you must keep your mouth closed if a monitor is watching you, for if the lizard sees your teeth, some disaster will surely befall you.

Iguanas

The largest lizards of the Americas are the iguanas. This group includes about thirty species of medium-to-large lizards and is part of a much larger family that includes such familiar species as the anoles, or American chameleons, the fence lizards, and the horned toads. The family today is almost exclusively New World, but the presence of a few primitive species on Madagascar point to a Gondwanaland origin. In Africa these must have been displaced very early by the agamids, which are in many ways their equivalents, and they may never have reached Europe, Asia, or Australia at all; at least they have left no fossil record. In South America, however, they became one of the dominant lizard families, and from there they spread to North America. Most surprising is the presence of two closely related iguana species on the Fiji and Tonga islands. Their nearest relatives are in Mexico and the southwestern United States. The best theory to account for their presence in the distant South Seas is to assume that a pair of iguanas, or perhaps a single gravid female, anticipated Thor Heyerdahl by several million years and made the trip by raft.[26]

The iguanas can chalk up several other successes in island colonizing. Two large species that approach a length of 5 feet inhabit the Galapagos. One of them, the marine iguana, is unique in being the only living lizard that has adapted to any extent to life in the sea. Although it spends most of its time on the lava rocks close to the coast, it feeds entirely on marine plants that it obtains by diving and scraping or pulling them off the rocks.[27] These iguanas can dive to at least 35 feet and stay submerged 15 to 30 minutes.[28] Darwin noticed that the lizards often squirted drops of fluid from their nostrils. This is the secretion of salt-excreting glands similar to those found in other marine reptiles and birds. He also noted that, although the lizards were good swimmers, they did not try to escape into the water, and if one was thrown into the sea, it promptly returned to land. He surmised that the lizards had no enemies on land but had an instinctive fear of sharks.[29]

The Galapagos land iguana was also very plentiful at the time of Darwin's visit, for he wrote, "I cannot give a more forcible proof of their numbers than by stating that when we were left at James Island, we could not for some time find a spot free from their burrows on which to pitch our single tent."[30] Today it is almost extinct, although the marine iguana remains in good numbers.

Another group of about a half-dozen large iguanas inhabits the southern Bahamas and West Indies. Although they do not grow to much more than 4 feet in length, they are heavy-bodied. As is true of most other large iguanas, they have a crest of flexible, spinelike projections down the middle of the back, a conspicuous dewlap, and large scales, or tubercles, on the head. In addition, the rhinoceros iguana has several stubby horns on its snout. All these iguanas are ground-dwelling animals that make their home in burrows or rock crevices. The common fate of large island reptiles has befallen them too; they are nearing extinction as man and his domestic animals take over their small world. On big islands, such as Cuba, Puerto Rico, and Hispanola, there are still remote, rough areas where the lizards can hold out. On small islands like Grand Cayman there is little hope for them.

The largest, or at any rate the longest, of all iguanas also happens to be the most widely distributed and abundant. This is the common iguana which has been reported to reach a length of 6 feet

7 inches,[31] although in most places an individual over 5 feet is exceptional. The species is found in lowlands from the states of Sinaloa and Veracruz in Mexico to southern Brazil and Paraguay and on some of the Caribbean islands. In parts of Central America, the name iguana is reserved for the female; the male is known as *gorrobo* or *guacho.*

The common iguana is a tree lizard, often climbing to heights of 40 to 50 feet. A single big tree may harbor a dozen or so adult lizards. One observer counted fifty-eight iguanas during a 3-mile drive in the Panama Canal Zone.[32] Iguanas are good swimmers and often drop into the water from overhanging limbs. They swim submerged and can remain under water several minutes. During cool or rainy weather they are more likely to be found on the ground.

Another type often seen in Mexico and Central America is the rock iguana. It does not belong in the giant category, for none of the five species grow much over 3 feet, but its habit of sunning on rock fences and its abundance around some of the Mayan temples and pyramids make it a conspicuous animal.

In the common iguana and most of the other large species, males reach a greater size than females, have proportionally larger heads, and have the crest of spinelike projections better developed. There are usually color differences as well. Males of the common iguana show yellow, orange, or reddish tints, while females are predominantly greenish.

Despite the abundance of the common iguana, its mating has never been described by scientists; the local people say that this occurs on the ground during the late fall and winter months. Iguanas usually nest in sandy places near water. A. Stanley Rand, a U. S. zoologist working in Panama, described a nesting aggregation of between 150 and 200 females on a tiny islet in Gatun Lake on the Canal Zone. Few if any of the lizards live there permanently; all swim out from the mainland during the winter dry season, mostly in February. After a good deal of exploratory scratching and random digging, the female begins work on her nest burrow, which may be as much as 6 feet long, although it is only about 12 to 18 inches below the surface. After completing the burrow, which often takes two days or so, she lays her eggs and fills the burrow with soil and plant debris. During this time she is very aggressive toward other iguanas. When the nest burrow is filled, she scratches up the ground and digs and fills shallow burrows until the entrance to her nest is effectively concealed. The young hatch early in the rainy season—late April or early May—and soon swim to the mainland.[33]

Communal nesting in burrows is characteristic of the marine iguana. The small amount of data available indicates that the other large iguanas also excavate nest burrows but do so singly.

All lizards of the iguana family and their Old World counterparts, the agamas, engage in characteristic head bobs, nods, and push-ups. In Moslem countries such as Pakistan, these lizards are particularly disliked because the people consider such movements a mockery of their prayers. A local name in Karachi for the common Indian garden lizard is *kafir-girgit* (infidel lizard), and it is particularly meritorious to kill one on Friday. These display patterns have now been analyzed for dozens of lizard species with the aid of movie film and found to be highly specific. They are performed by adults of both sexes, as well as by young just out of the egg. The displays of tree lizards are generally more elaborate than those of species that live on the ground.[34] Their chief function seems to be one of recognition, but possibly other messages can be conveyed.

The common iguana is herbivorous as an adult. One of its favorite foods in Panama is a type of wild plum; one lizard had seventy-nine of these fruits in its stomach. They eat many other types of fruits, leaves, blossoms, and buds. The young feed mostly on insects.[35] The rhinoceros iguana and the other big West Indian species are omnivorous, feeding on birds, small mammals, and lizards, as well as a variety of plant materials.

Large iguanas defend themselves in the same way as monitors, by lashing with the tail, biting, and clawing, although they are generally better-tempered. One herpetologist reported a blow from an iguana's tail which raised a welt that lasted several hours.[36]

In many part of Latin America the common iguana is a delicacy, sometimes known as *gallina de palo*, or chicken of the tree. Large gravid females are the best eating, for the eggs can be cooked and added to the meat. In Panama the meat is said to be most flavorful in September and October; in Mexico during March and April. This lizard is a convenient source of meat in the tropics, for if it is not overheated it can be kept alive a long time without food or water. In Central American markets it is common practice to fold the lizard's legs over its back, slit the toes, and put the claws of one foot through the tendons of the other. Other species of iguanas are also eaten. Some of the sailors who replenished their larders on the Galapagos considered the land iguana better eating than the tortoises.[37]

Paul Swanson, a horticulturist who has pursued herpetology as an avocation in many parts of the world, says that in El Salvador it is widely believed that persons contract warts from the blood of iguanas and that one person can transmit them to other persons by contact with his urine. They ridiculed the idea of contracting warts from toads. He also tells of a Panamanian ventriloquist who sold a "talking iguana" to a gullible Yankee tourist.[38]

11. Snakes: The Big Five and Their Relatives

It is sometimes hard to tell where tales of purely mythical giant snakes end and those of presumably real giant snakes begin. The Midgard Snake that encircles the earth, and Yulungurr, the great python of the Australian bushmen whose body spans the heavens like a rainbow, are clearly symbolic animals having no more real existence than the British lion or the Russian bear. Mythical surely, if not symbolical, is the story told by the Indians of British Guiana, who say that a group of their people were making their way through dense jungle when they came to a smooth wall too high to climb. After following it two days, they found it ended in the head of a gigantic anaconda, and they fled lest they arouse the sleeping monster.[1]

When Alexander the Great's soldiers told of snakes in India that had eyes as large as shields and were able to swallow oxen and stags, they were unquestionably speaking of real animals, probably the Indian python and the crocodilians of the Indus. But medieval scholars never questioned that India and Ethiopia bred snakes large enough to swallow elephants, and Chinese literature mentions the Pa snake which could accomplish the same feat, ejecting the bones after a period of three years.[2]

Even today, people will believe almost anything about snakes and accept almost any figure for the maximum length they reach. A Brazilian newspaper article in 1948 told how a detachment of the army had to be called to dispatch an anaconda 156 feet long that knocked down buildings and upset automobiles before it was finally killed. This account was repeated in a 1971 magazine story along with a report of a 131-foot anaconda illustrated on a picture postcard. The photographer was doubtless inspired by the bronco-sized jack rabbits often shown on post cards from Texas.[3]

Captain C. R. S. Pitman, a British naturalist and authority on African snakes, was told in all seriousness of a 130-foot python that inhabited a pool in the Bwamba escarpment of Uganda. It was said to be of prodigious age and venerated by the local people; however, they eventually killed and ate it. Pitman's informant obtained "the mouth with teeth and two pieces of its vertebra," but these apparently never reached a scientific collection. Regarding the snake's length, the informant said, "I have no reason to disbelieve what the headman says: he is a reliable man and he measured it with a linear tape measure . . . every single soul who was present when the snake was measured states that it much exceeded the tape measure. . . . I was shown the place where they stretched it out . . . this was certainly approximately 130 feet."[4] Stories of such snakes remind me of an old saying from my ROTC days about the Springfield rifle: "It weighs 9.8 pounds. After you've carried it a couple of miles, the decimal point drops out."

Leaving aside for the moment the purely mythical serpents and the sea monsters whose exegesis is a specialty in itself, what are the giant snakes of the world and what size do they attain? There are five truly giant snakes that are known to reach lengths of 20 feet or more: the anaconda, reticulated python, African python, Indian python, and amethystine python. A dozen or so additional species fall in the 10- to 19-foot range. All the rest of the 2,700-odd species of living snakes are under the 10-foot mark, although a few are so heavy that they outweigh some of these that reach lengths greater than 10 feet.

The snakes are the most recent major development in reptile

evolution. They are an offshoot of a lizard lineage that literally went underground sometime during the Cretaceous. The oldest fossil unquestionably that of a snake was found in Patagonia and represents an animal about 6 feet long. It is more lizardlike than any living snake but shows no close relationship to any group of existing lizards. Its nearest relatives among the snakes are the pipe snakes that live today in tropical America and southeastern Asia, but recent investigation shows that this relationship is not so close as was formerly believed.[5]

Large size is an attribute which snakes apparently developed early in their history, and all the largest snakes today are members of the archaic family Boidae. This family flourished about 50 million years ago during the Eocene, and some of its members, such as *Gigantophis* of Egypt and *Madtsoia* of Patagonia, reached lengths of about 33 feet.[6] Assuming that these fossils were of average-sized rather than exceptionally large individuals, these snakes must have been bigger than any present-day boas or pythons. At about the same time there were some very large, apparently marine snakelike creatures around. Their remains are so fragmentary that the size cannot be reliably estimated, and there is some question whether they were really snakes. *Palaeophis*, whose fossils have been found in Europe and the United States, may have been a very elongate mosasaur, and it has been suggested that *Archaeophis* was a giant eel rather than a snake.[7] The survival of a few of these creatures down to recent times has been suggested as a factual basis for some of the sea-serpent stories, but it seems totally unlikely that such a creature, which would be air-breathing and probably an inhabitant of comparatively shallow waters, could survive undetected by science.

The division of the family Boidae into boas and pythons is based on certain skeletal features, such as the presence in pythons of a separate supraorbital bone and premaxillary teeth, and one interesting biological feature—all boas, so far as known, are live-bearing, while all pythons lay eggs. Boas are found in both the New and the Old World, while all pythons are Old World snakes. For a long time, *Loxocemus*, a moderate-sized shiny brown snake found from western Mexico to Costa Rica, was believed to be a python and was referred to in popular books as the Mexican dwarf python, or Colima python. Recent work indicates that it is not a python, although there is some question as to just where it does fit in the scheme of snake classification. Most evidence suggests that it is a primitive species with no close living relatives. The boas and pythons together form a rather heterogeneous family. The small West Indian boas of the genus *Tropidophis* and the even smaller rubber boa and rosy boa of the western United States have no more in common with the boa constrictor than that species does with the Indian python. The little burrowing Calabar python of West Africa is another highly aberrant species, very different from the larger pythons. This snake, like the Colima python, may be a remnant of an almost vanished group of primitive pythonlike snakes.

In addition to being an ancient family of snakes, the boas and pythons seem to be one that is declining. The family contains about seventy-five living species, and many of them are either rare or very limited geographically or both. Most of the biggest species, however, have extensive ranges and, until reduced in numbers by twentieth-century man, were fairly plentiful. In all tropical regions of the world, there seems to be room for one or two species of really big snakes; however, the ecological niches for such reptiles are stringently limited. Any snake of 15 feet or more is of necessity a master predator and must take its cut from near the top of the food pyramid. Here it is in competition with large carnivorous mammals, the very largest raptorial birds, and the crocodilians. All of these are occasionally foes as well as competitors, so a giant snake's life is somewhat chancy even when it is well grown, not to mention the period when it is a tasty unprotected infant. The smaller species of boas and pythons have found it hard to compete successfully with the more recently evolved snakes. Most of the modern snakes fall into a huge, untidy family that most herpetologists still call the Colubridae, although a few brave souls have tried to split it up into natural subdivisions. The colubrid snakes, substantially assisted by the venomous elapids and vipers, have tended to push the smaller boas and pythons into restricted niches such as islands, deserts, and cloud forests.[8] Why this should be so is not clear. Perhaps the colubrids and other modern snakes are a little faster in their reactions, reach maturity sooner, are better able to withstand rapid changes in temperature, and utilize available food supplies better than their conservative ancestors. Nevertheless, the boas and pythons have managed to keep a stranglehold on the niche for giant species.

The Anaconda

Two American herpetologists, James A. Oliver, director emeritus of the American Museum of Natural History and director of the New York City Aquarium, and Clifford H. Pope, formerly on the staff of the Field Museum of Natural History, have provided excellent critical evaluations and summaries of the records of maximum length attained by the giant snakes. They agree that when both length and bulk are considered, the largest snake in the world is unquestionably the anaconda. This enormous boa is found in streamside and swamp habitats in South America east of the Andes from Colombia and Venezuela to northern Bolivia and south central Brazil. There are several apparently reliable records of anacondas more than 30 feet long. A party of oil prospectors working in the region of the upper Orinoco River in eastern Colombia shot an anaconda that was 37.5 feet long by measurement with a surveyor's tape. After measuring the snake, they left it for a short time and returned to their work. When they returned to skin it, the reptile was gone. Evidently it had recovered enough to crawl away. Fredrico Medem, a Colombian herpetologist and expert on crocodilians, saw an anaconda which he estimated as between 30 and 40 feet long and obtained a report of a specimen 10 meters, 25 centimeters (about 34 feet) long that was killed on the Rio Guaviare in Colombia in 1956. General Candido Mariano de Silva Rondon of the Brazilian army, a reliable explorer and naturalist, measured an anaconda slightly over 38 feet long that had been killed by Indians in south central Brazil. Afranio do Amaral, a herpetologist long associated with Brazil's Instituto Butantan, which is famous for research on venoms of snakes and other poisonous animals, assembled numerous reports of very large anacondas; the biggest he considered reliable being one of a snake 11 meters, 28 centimeters (slightly over 38 feet). Vincent Roth, former director of the British Guiana Museum, on the Barima River in 1921, shot an anaconda which he estimated as having a length of 34 feet.[9] All these are actual measurements made by presumably reliable persons, or the estimates of naturalists familiar with large snakes in the field. There are numerous less well authenticated accounts, ranging up to an estimate of 62 feet for a specimen killed by the English explorer Major Percy Fawcett on the Rio Abuna near the Brazil-Bolivia border in 1907.[10] Unfortunately, all these encounters took place in the remote hinterland of South America, and no part of the snakes in question was saved to support the story. To the best of our knowledge, no anaconda over 25 feet long has ever reached a zoo or museum in the United States or Europe.

For all its impressive size, the anaconda is little known, and there is even doubt whether it constitutes one species or several. The latest catalogue of South American snakes indicates that there is one subspecies in the Amazon drainage and another in northern South America. In 1936, two species were described from Marajo

Island at the mouth of the Amazon, but they may be only variants of the widespread Amazon form.[11] Further south in South America, chiefly in the drainage basins of the Paraguay and Parana rivers, there lives the smaller southern anaconda, whose maximum length does not exceed 12 feet.

The name anaconda is of Sinhalese origin and was originally applied to the Indian python. The Portuguese evidently brought the name from Ceylon to Brazil, where it became associated with the largest snake of the New World. Aboma is another name used for the anaconda in some of the older literature. This is an African name also originally applied to a python but brought to America by Negro slaves and given to a different snake. The anaconda also has several names in South American Indian dialects. *Sucuri* is a common name in Brazil, and *camudi* in Guiana.

Like all the big constricting snakes, and like the great majority of snakes large and small in the tropics, the anaconda is most active at night. However, in several accounts of encounters with giant anacondas, the snake was found in the open during the day and appeared to have recently swallowed a large meal. Perhaps after such large feedings snakes seek the sun's warmth to help speed digestion. On the other hand, they may simply have been too gorged to seek their usual diurnal shelter.

The Reticulated Python

The late Raymond L. Ditmars, formerly of the Bronx Zoo, whose popular books on snakes greatly influenced our generation of herpetologists, considered the reticulated python the largest of snakes. He gave the record length as 33 feet. This is probably based on an account by the late Henry C. Raven, an American museum collector who between 1912 and 1923 traveled extensively in what is now Indonesia. Raven saw a very poor photograph of a python of that size which had been killed at a mining camp on the island of Celebes. The snake had been measured with a surveying tape, apparently with some care. Henry Trefflich, an American animal dealer, is said to have obtained a reticulated python 32 feet long from an unspecified source. If the measurement was accurate, this seems to be the largest snake ever taken alive. There is a report of a 33-foot reticulated python killed in Java and one of a 30-foot specimen killed near Penang, Malaya, in 1844, but the original sources of these accounts are unknown.[12] Major Flower mentions a 27-foot Malayan specimen whose skin was stretched to a length of 33 feet.[13] A few pythons of 28 to 29 feet have been exhibited in zoos. "Colossus," a 28.5- foot specimen, died of old age and pneumonia at Pittsburgh's Highland Park Zoo in 1966, at an estimated age of 30 years. It was reported to weigh 300 pounds.

Measuring a live snake of any appreciable size is not easy; we have tried several times with specimens in the 6- to 8-foot range and had difficulty in getting three consecutive measurements that agreed within an inch or two. The difficulties are compounded manyfold with huge snakes of 20 feet or more. It is quite impossible to straighten out such a reptile without general anesthesia (which is dangerous for snakes), so measurements are taken by running a string along the middle of the snake's back and measuring the string. This method is also subject to considerable error.

Very large reticulated pythons (25 feet or more) reach zoos and other scientific collections more frequently than do anacondas of comparable size. This is doubtless due to the python's range, which is extensive in accessible coastal areas, and to its ability to adapt to living in densely populated regions. It occurs from southern Burma, Thailand, and the peninsula of Indochina through Malaya and nearly all the islands of Indonesia northward to the Philippines.

Major Flower's account of the species in Bangkok at about the turn of the century is worth repeating:

> Strange to say it is not in the quiet jungle forest that the python seems to prefer to live but in the busiest spots along the Menam, where steamers and junks are loading and unloading, steamer-launches whistling, steamsaws buzzing, rice-mill chimneys filling the air with smoke, and hundreds of noisy coolies passing to and fro; here he selects some hole or crevice in building, timber-stock, or bank to spend the day in, and at night makes an easy living devouring fowls, ducks, cats, dogs, and, it is said, pigs.[14]

He mentions a 20-foot specimen found under the floor of a stable in the middle of the city. Malcolm Smith, who lived in Bangkok a few decades later as physician to the Royal Court of Siam, reported the snakes still regularly found in the city.[15] They sometimes entered the gardens of the royal palace, where one of them ate a Siamese cat belonging to the king's family. A taxi driver in Legaspi City on Luzon was recently reported to have run over a 21-foot python with his cab and then finished it off with a boulder. Simon Campden-Main, an American herpetologist who served with the armed forces in South Vietnam, reported the reticulated python to be common throughout that country, often in populous areas and nearly always near water.[16]

Pythons have been popular as pets with American personnel in Vietnam. In 1971, an air-base dentist and a veterinarian kept an 8-foot specimen at Bhu Cat, where they exercised it on weekends. The 20th Engineer Brigade at Bien Hoa had a 12-foot female python which was called Moonbeam. And sailors at the U. S. Naval Base post in Can Tho in the Mekong delta had a python named Sam who reportedly grew from 8 to 16 feet in 3 years on a diet of two live chickens a month.[17]

A veteran helicopter crewman testified that at his headquarters in Can Tho a python was regularly used as an incentive to make Vietcong prisoners and suspected sympathizers talk. He stated in February 1971:

> . . . they had an eight foot python snake which was kept at the camp in a cage, supposedly for rat control. When we had prisoners or detainees who were brought to the B. team, they were immediately questioned, and if they balked at all or sounded like they weren't going to be cooperative, they were simply placed in a room overnight. This was like a detention room; the door was locked, and this snake was thrown in there with them. Now the python is a constrictor, similar to a boa . . . it probably can't kill a full-grown American or a large male, but it sure terrified the Vietnamese. Two of them usually were in a room overnight with the python snake, struggling with it most of the night, I guess, and we could hear them screaming. In fact, on one instance, they had to go in there and gag the prisoners, so they wouldn't keep everyone awake all night. In the morning they were usually more cooperative.[18]

The reticulated python is a good swimmer and sometimes enters the sea. It was one of the first land animals to return to Krakatoa, the small island in the Sunda Strait between Java and Sumatra that was virtually sterilized by a colossal volcanic eruption in 1883.[19] Reticulated pythons also seem to be good climbers, although accounts of this species and other large snakes dropping from trees onto their prey seem to be completely unsupported.

The African Python

Third in size among the giant snakes is the African python, which inhabits most of that continent south of the Sahara, occurring in the equatorial highlands to about 7,000 feet. The largest specimens reported are from West Africa. One measuring 9.8 meters (about 32 feet) was shot near Bingerville, Ivory Coast, in 1932. Another from Adiopodoume in the same country measured 7.5 meters (about 24 feet).[20] One whose skin stretched to 30 feet was killed on the Ngeri Ngeri River in what is now Tanzania, but specimens of 20 feet or more are extremely rare in East and South Africa.[21]

The African python is an ecologically adaptable species reported from many kinds of terrain, ranging from rain forest to semidesert. It is most at home in open grassland near water and, except for the anaconda, seems to be the most aquatic of the giant snakes. It is sometimes found near human habitations; the record 32-foot specimen was found in a bougainvillea hedge surrounding a school.

There is no evidence that the African python inhabited Egypt or any other country of the Mediterranean littoral, so the story of the death of the Trojan priest Laocoon and his sons from the attack of giant serpents is probably pure fabrication. However, the snakes in the famous Laocoon sculpture group (ca. 150 B.C.) have the head and body conformation of pythons. Similarly, Virgil, in Book V of the *Aeneid* (ca. 20 B.C.), describes a great snake that encircled the tomb of Anchises with "huge seven coils, in each seven folds . . . dark blue streaks its back lit up, its scales a sheen of spotted gold as (when the sun shines opposite) the bow darts from the clouds a thousand varied hues." This iridescence is characteristic of pythons, especially when they have just cast their skins. Pliny wrote: "We see in Italy other serpents . . . so big and huge that in the days of the Emperor Claudius [A.D. 41–54] there was one of them killed in the Vatican, within the belly whereof there was found an infant all whole."[22] This snake must have been an escapee from the imperial menagerie. Clearly the people of classical Greece and Rome were familiar with some species of python, probably the African which still occurs in Sudan and which could have been shipped down the Nile to Alexandria and thence to Rome or Athens. They also undoubtedly knew of the Indian python through hearsay and may well have seen skins or even living specimens transported in trade from the East.

The Indian Python

A pleasing pattern and a reasonably tractable disposition have made the Indian python a favorite of showmen in both the Eastern and Western worlds, and it has probably been seen by more people than any other giant snake. If reaching a length of 20 feet or more is essential to qualify a snake as a giant, the Indian python only questionably makes it. The biggest specimen reliably recorded is one shot in Assam by the Maharajah of Cooch Behar about 1875. It was 19 feet 2 inches long and weighed 200 pounds.[23] Since there are a comparatively large number of records of specimens between 18 and 19 feet, nearly everyone who has written about this snake assumes that there were or still are 20-footers. Nearchus, admiral of Alexander the Great, was probably the first European to describe this python. He gave its length as 16 cubits (about 24 feet), which may have been only a slight exaggeration.[24]

The Indian python is found from the Sind-Baluchistan border in West Pakistan throughout India and Ceylon exclusive of the higher mountains and eastward through southern Asia to the province of Fukien in southeastern China. However, it evidently does not occur on the Malay peninsula, nor has it the island-colonizing tendency so evident in the reticulated python. There are scattered records for Indonesia, even as far out as Celebes, but these seem to be based on waifs or possibly small populations originating from individuals introduced by snake charmers. The populations of this snake from the southeastern part of the range are considered a different subspecies, often known as the Burmese python. This form averages slightly larger than the typical Indian subspecies, although there is little difference in maximum sizes recorded.

Over a good part of southeastern Asia this snake coexists with the reticulated python and evidently occupies much the same habitat. This is a unique situation and one that would repay study to determine how two large constricting snakes can live together without a degree of competition that would normally be expected to oust one or the other. The anaconda and the boa constrictor coexist over a large area in the American tropics but they do not really compete, because the anaconda exploits a semiaquatic habitat while the boa constrictor is terrestrial and arboreal. This does not seem to be the case with the two big Asian pythons, but no one has investigated the matter thoroughly. Both these pythons are relatively versatile snakes that are at home on the ground, in the trees, or in the water. The reticulated python seems to be more of a climber than the Indian python, but both have been reported in trees with some frequency.

In West Pakistan the Indian python is largely restricted to that part of the Indus valley that has dense undergrowth and is south of the limit of killing winter frost. There evidently are a few pythons in the more arid Hab River valley, but we were never able to obtain a specimen from this locality. Pythons in the Indus valley are usually found close to water and often actually in it. A *zamindar*, or land-owner, in the Sanghar District of Sind told me that as he was inspecting his fields, he noticed a small irrigation culvert which was blocked up. He asked a villager who was with him to clear it out, but the obstruction proved to be a fair-sized python that caught the man's hand between its coils. "He tried to say, '*Samp, sahib*! [snake, sir!],' but he was so frightened he could only sizzle. Afterward he developed a fever and took to his bed for a fortnight." In the same district I saw a python about 10 feet long that had been gaffed by fishermen in the East Nara. It had been badly injured and a large chunk of its head was torn away, but it was alive and seemingly in no great distress.

In the western part of the Indus delta are little rocky knolls that rise abruptly from the floodplain and often are capped by ancient stone tombs or watch towers. Pythons sometimes frequent these knolls, living in rock crevices or in the ruins themselves. In January 1962, snake hunters brought in from this area four very fat pythons, the largest 12 feet long, that they had dug out of burrows in which the snakes characteristically hibernate from late October through February.

The largest python that was brought to the Karachi Zoo during our stay in Pakistan was a 15-foot specimen weighing 115 pounds. It also was dug out of a burrow during the winter. The hunters' account as reported in *Dawn*, a local English newspaper, reads:

> My child Rasul Bux was ill and I had no money to buy medicine. The crop had failed and one of the bullocks was pulled down at dusk by desert wolves. That night, my wife, weighed with grief at the misfortune, sobbed to me, "Dear, why don't you get some money? . . ."
>
> After consoling her, I prayed to God and slept in the biting cold which kept reminding me that unless some money came in a hurry, my family would perish. . . . Maluk, my neighbor had recently sold an untanned python skin in Hyderabad for

> Rs. 50/-. It occurred to me that if, by risking my life, I could save the family at this critical juncture, it would be well worth it.

The actual hunt and capture was pure anticlimax.

> We had little difficulty tracing the python's home . . . it did not resist much; but it was heavy as could be. On our way to Karachi, we are not sure it enjoyed its first journey in a train! We will now, Sir, beg leave of you, because my son will be hoping that his daddy will bring medicines and lollipops from the city.[25]

The Amethystine Python

Least known of the giant snakes is the amethystine python, which is the largest of the dozen or so python species that inhabit Australia, New Guinea, and adjacent islands. Such a concentration of pythons, all of them reaching lengths of 6 feet or more, is unknown anywhere else. It may partly be explained by the scarcity of mammalian predators and the virtual absence of terrestrial colubrid snakes in this part of the world. These circumstances considerably reduce competition for food and habitat niches, although the venomous elapid snakes have also proliferated in this region and occupy many of the niches held by colubrid snakes in other sections of the world.

The amethystine python is known from northeastern Australia, New Guinea, and many of the eastern islands of Indonesia, such as Ceram, Halmahera, and Timor. It is a more slender snake than the other big pythons. The largest specimen on record is one

28 feet long killed at Greenhill near Cairns, Australia, in 1948. There is another record of an Australian specimen of 23 feet 8 inches, but most adults are 10 to 15 feet.[26] Almost nothing is recorded concerning the size reached in other parts of the range. Men who served with the armed forces in New Guinea during World War II have told us of encountering pythons of 20 feet or so on that island, but these were estimates that have to be treated accordingly.

Eric Worrell, an Australian herpetologist, is almost the only scientist to have reported field experiences with this snake. A collecting party that he headed took thirteen specimens in wild, rocky gorges at the junction of the Johnstone and Beatrice rivers in northeastern Queensland. The snakes were found among the rocks and not in the water. The species is also reported from other types of habitat but nearly always in scrub or jungle. It is said to be a good climber.[27]

The Boa Constrictor and the Smaller Pythons

Although the boa constrictor is widely considered the prototype of a giant snake, the reptile to which this name is properly applied falls short of the 20-foot mark and over most of its range rarely exceeds 10 feet. Much of the confusion has come from applying the name to the big pythons of Africa and Asia. Moreover, early explorers of the American tropics did not always distinguish between the anaconda and the boa constrictor, although the two snakes are quite different and are always distinguished by aboriginal peoples whose lands they inhabit. The name boa, incidentally, is of Latin origin and according to some means "water snake."

Others derive it from *bos* (cow) and say that the snake "pursues herds . . . and clings to their udders . . . it destroys them by suckling."[28] To add to the confusion, herpetologists are not in agreement as to the proper application of *Boa* as a scientific name. Some apply it to the genus that includes the snake popularly known as boa constrictor, while others say it should be used for a genus of smaller arboreal boas that includes the beautifully colored emerald tree boa of South America. This is one of the instances where a scientific name has been more troublesome than helpful. Among contemporary zoologists, the popular name boa constrictor is more meaningful than the scientific name.

Be that as it may, the boa constrictor is a distinctive snake and has been recognized for a long time. It has a very extensive range, being found in lowlands from northern Argentina northward to the Mexican states of Sonora and Tamaulipas, both of which border the United States. An easy day's drive from Brownsville, Texas, or Nogales, Arizona will put you in boa-constrictor country. It also inhabits some islands of the West Indies and islands off the coast of Mexico and Central America. It is quite a variable snake, and eight subspecies are currently recognized.[29] The variety most in demand in the pet trade and for exhibition is a warm tan to amber with large, rich brown blotches that become reddish toward the tail. This is the form typically found in Amazonian South America and northward to the Caribbean coast. Boas of Mexico and northern Central America tend to be more grayish with less contrast in color and are generally less attractive. Moreover, they have the reputation of being bad-tempered. A heavy-bodied, dark variety of boa constrictor is found in Argentina and Paraguay, while in southern Brazil there lives a subspecies with a pale ground color and small blotches. Boa constrictors with very little pattern are found in the semi-arid country of northwestern Peru and on the island of Taboga off Panama.

The largest boa constrictor on record was killed in the central range of Trinidad by the English biologist Colin S. Pittendrigh while he was doing malaria control work on that island during World War II. It measured 18.5 feet and is by far the biggest specimen reliably reported.[30] In northern South America and southern Central America, where the species apparently reaches its greatest size, 12-foot specimens are exceptional, and the record is about 15 feet. The southern subspecies and those inhabiting the West Indies reach maximum lengths of 8 to 10 feet.

As is usually the case with a wide-ranging animal, boa constrictors adapt to various habitats. This snake's body is somewhat compressed laterally and flat on the belly, rather like a loaf of bread in cross section, a conformation usually associated with arboreal tendencies. Over much of its range the boa constrictor is an inhabitant of tropical forest and seems to spend a good part of its life in trees. In the northernmost part of its range in western Mexico and in northern Peru west of the Andes, it invades a region of semi-arid scrub. In the thorny acacia forest of the Mexican state of Sinaloa it is the commonest large snake, although its maximum length reaches only about 7 feet. Here the snakes seem to be more terrestrial, living in burrows and prowling in the open after dark.[31] Boas also can adapt to living around towns and villages where they find plentiful food in the form of rats, chickens, and probably an occasional cat or puppy. We saw a 7-footer killed on the road just outside the Mexican city of Ciudad Mante. Boas used to appear fairly regularly as stowaways in banana shipments, along with a whole assortment of fascinating tropical snakes and other animals. We got our first boa from a fruit market in Franklin, Indiana. Soon afterward we became acquainted with Thomas Payne, a most interesting old man who worked for a wholesale fruit dealer in Indianapolis. Once or twice a month for several years he would show up at the medical school with a snake, mouse opossum, scorpion, spider, or other intriguing and exotic creature that had arrived in a shipment from tropical America. Twice workers at the fruit markets were bitten by reptilian stowaways—one an eyelash viper, whose bite can be serious, and the other a cat-eyed snake which is rear-fanged but practically harmless to man. Today, bananas are treated with so many sprays and dips and so effectively packaged that very few animals survive, and northern biologists are deprived of an intriguing glimpse of the neotropical biota.

As a result of the current popularity of boa constrictors as pets, many have escaped and become temporary residents of American cities, but none seems to have survived long. Even in a rural area with plenty of cover, in most parts of the country an escaped boa constrictor would probably be killed by the first hard freeze. Only in southern Florida and perhaps in the lower Rio Grande valley is there a good chance that a boa could maintain itself indefinitely.

In addition to the boa constrictor and the anaconda, about thirty other species of boas live in the American tropics, quite a few of them on the islands of the West Indies. By far the largest of these is the Cuban boa which has been reported to reach lengths of 12 to 14 feet.[32] It is found in a variety of situations, including forests, cane fields, swamps, and beaches. Several accounts mention its fondness for living in caves and feeding on bats.

Another concentration of moderately large constrictors—pythons this time—occurs in the northern Australia–New Guinea area. Best known of these snakes because of its wide range and attractive pattern is the carpet python which occurs over much of Australia, exclusive of the highly arid interior desert. In the southeastern part of the country it is replaced by the closely related diamond python, another ornamentally marked reptile. Both these snakes live in a variety of habitats, often around farms and towns. In limestone areas in the far northern part of Australia is found the drab-colored olive python. In the same general area, the water python frequents streams and estuaries, sometimes living in mangrove swamps where the water is quite salt.[33] It also occurs on New Guinea and islands of the Torres Straits. On the main island of New Guinea and some of the associated smaller islands lives the little-known Papuan python. All these pythons are reported to reach lengths of 10 to 14 feet, although they average smaller.[34] Several species in the 5- to 9-foot range also live in this region.

12. Biology and Behavior of Boas and Pythons

Reproduction

Sex is easier to determine in boas and pythons than in most kinds of snakes. One of the primitive characteristics of this family is the presence of a pair of spurs near the vent, which represent vestiges of hind limbs and are considerably larger in males than in females. They are conspicuous enough that they figure in regional folklore. In South Africa the python is said to push them into the nostrils of its victim while constricting.[1] The Malagasy boa is reputed to make wounds with its spurs and suck the blood from them.[2] In West Africa, the natives believe that the spurs are venomous, but an eighteenth-century Dutch trader commenting on the python's habits said that the "two claws . . . helped him [the snake] to erect himself and contributed to his more expeditious march."[3]

In spite of this helpful feature, little is recorded of differences in size between the sexes among the very large snakes. What information is avilable indicates that females reach a greater size than males. The huge reticulated python "Colossus" was a female, and female Indian pythons are said to be larger than males. No marked difference in length between the sexes has been reported for the African python, but females are heavier. There is a general impression among zoo men and animal dealers that female boa constrictors average larger than males, but this is partly contradicted by a report of a boa only 3 feet 11 inches long that gave birth to four healthy young.[4]

The courtship and mating of large boas and pythons in the wild has very rarely been reported; however, there is a fair amount of information on matings of these reptiles in captivity. There is also a great deal known about the mating behavior of smaller snakes of other families, but it is unwise to assume that their behavior is similar to that of the giant snakes except in a very broad sense. There is good reason to think that the sexes find each other by scent and that the male is more active in seeking the female than vice versa. Snakes of both sexes have a pair of elongate glands in the base of the tail with openings at the edge of the vent. These are full of a creamy, strong-smelling material, and even the comparatively insensitive human nose can detect differences in odor between species. Since the contents of these glands are often ejected freely when the snake is angry or frightened, some herpetologists believe their function is defensive. Others think that the defensive function is secondary and the principal function of the secretion is to enable the sexes to find each other for mating. However, no one has reported either an increase in the amount of secretion or a change in its character during the breeding season. Moreover, some observers have found that the skin odor of the female is more attractive to the male than is the secretion of the scent glands.[5]

The close-range courtship behavior of boas and pythons involves a good deal of tongue-flicking by the male and jerky, pushing movements, both being characteristic of many other snake species. Males of many lizards and snakes grasp the neck or body of the female in their jaws during courtship and copulation, but this has not been reported for the giant snakes. A unique feature of courtship in the boas and pythons is the scratching or tickling of the female's cloacal region with the male's spurs, which are held at almost right angles to his body. This may continue for as long as a few hours and usually leads to the female's raising her tail and bringing her cloaca into apposition to that of the male.[6] This facilitates the insertion of the penis.

Only snakes and lizards possess a double penis which retracts into the base of the tail and is in no way associated with the passage of urine. Early anatomists thought both organs were used simultaneously and referred to them as "hemipenes." The name has stuck, but each organ functions independently of the other, and only one is used in any single act of coitus. In snakes, and to a lesser degree in lizards, the penis comes in an amazing variety of shapes and sizes, being absolutely enormous in the king cobra and so tiny and slender in the Afro-Asian sand snake that it is hard to see how it can have any function. Moreover, in many snakes it is ornamented with elaborate arrangements of cups, flounces, and spines, the latter sometimes being so large and numerous that coition would seem to be uncomfortable and fraught with danger for both parties. The significance of all this complicated equipment is speculative. It may discourage hybrid matings or it may keep the genitalia in position so that sperm can be deposited in that part of the female reproductive tract which insures maximum efficiency in fertilizing the eggs. It also helps to keep the pair together through a lengthy period of copulation; for example, rattlesnakes have been reported to couple for almost 23 hours, and pythons from 35 minutes to 3 hours.[7]

Pythons are egg-layers and choose nesting sites much like those used by smaller snakes—hollow logs, mammal burrows, and cavities under rocks and around the roots of plants. Occasionally they nest in high grass and other comparatively open places. For the African python, Indian python, and reticulated python, average clutches contain 30 to 50 eggs and exceptionally large ones up to about 100. The eggs are oval, 3 to 4.5 inches in greatest diameter, and have flexible, parchmentlike shells. The brooding habit is well developed in these pythons, the female gathering her eggs into a compact mass and surrounding them with her coils in such a way that the eggs are almost completely hidden yet not subjected to harmful pressure. The female's head rests on or near the top of the egg pile. Indian and African pythons brooding eggs in captivity are largely oblivious to disturbance by scientists measuring their body temperature and photographers taking their pictures; reticulated pythons are reported to be more touchy and occasionally strike when approached. No one has determined the snakes' reaction to a natural egg thief such as a mongoose or monitor lizard. Brooding pythons remain with their eggs throughout the incubation period of 8 to 13 weeks, although they may leave them briefly to eat, drink, or shed their skins.[8]

Brooding is a fairly common trait among snakes. In addition to the pythons, it has been reported for several venomous species, such as the Asian cobra, the king cobra, and some of the kraits and Asian pit vipers. The mud snake, a nonvenomous species of the southeastern United States that is prominent in folklore for stinging with its tail and rolling like a hoop, regularly broods its eggs, and some other North American species are said to do so occasionally. It is sometimes hard to decide whether a female is really brooding her eggs or simply remaining near them because the nest also happens to be a good hiding place. We once found an adult female milk snake a few inches from a clutch of eggs nearly ready to hatch, but there was no way to be certain that these were her eggs or that she had been with them continually.

The Indian python does more than simply brood her eggs; she also incubates them. This was first observed in 1832 in the poorly heated reptile house of the Jardin des Plantes in Paris where a brooding female allegedly raised her temperature to 106.7° F. and later held it at about 90°. Other observers reported similar findings but did not always consider the possibility of heat transfer from the substrate or other external sources. Careful observations on a brooding Indian python in the Bronx Zoo showed that she could raise her temperature as much as 4.7°C. (ca. 8.46° F.) above that of her environment and keep a remarkably stable temperature of about 90° F. The female does this by spasmodic muscular contractions similar to shivering in mammals. These begin when the temperature drops below 32°C. and reach a rate of about thirty per minute. One poor female literally shook herself to death in a vain effort to keep her eggs properly warmed. Another female showed false brooding behavior with a temperature rise over a period of several months. This may have been a symptom of illness, for she was suffering from a respiratory infection that proved fatal. There is a report of similar shivering in the blood python, a small but stout species of southeastern Asia. African pythons have been reported to be able to keep their eggs slightly above ambient temperature, but shivering has not been observed.[9] It is curious that this temperature-elevating behavior is so much more highly developed in the Indian python, which ranges no farther north of the equator than the African python does to the south and lives at no greater elevations. Moreover, in the Indus valley and presumably in other northern parts of the range, evidence indicates that the eggs are laid during June and July, a season when a cooling system would seem to be of more advantage than one for raising temperature. Perhaps the species originally inhabited a cooler region than the one it lives in today.

Live-bearing, which is characteristic of boas, has some advantages and some disadvantages as compared with egg-laying. The mother snake's body is a living incubator in which moisture is optimal and temperature is easily controlled by her movements. The developing embryos are safe from fungi, insects, and many small predatory mammals and reptiles that might devour an egg but would be reluctant to attack an adult snake. On the other hand, if a female blue racer is squashed by a car five minutes after she has laid her eggs, her brood has the same chance for survival that it would have were she alive. The brood of a live-bearing copperhead share the risks of their mother from conception to birth. Wilfred Neill says that live-bearing in snakes has developed from the primitive egg-laying state under three sets of circumstances. One is in highly aquatic species, where going ashore to seek a suitable nesting site would impose special hazards on the female and comparable hazards on the hatchlings seeking to return to the water. The second is in adaptation to a cold climate, where eggs in a nest are at the mercy of weather while a gravid female can follow the sun's warmth and hibernate with her developing brood if necessary. The third is among snakes that are well defended by reason of being large or venomous or both.[10]

Once live-bearing has become established as a mode of reproduction, a return to egg-laying seems to be almost impossible, so the habit has been retained by the small boas. It probably helps one species, the rubber boa, to range well above the Canadian border in British Columbia. The only snakes that range appreciably farther north in western North America, are three species of garter snakes, which are also live-bearing. The anaconda and boa constrictor produce almost as many live young as the larger pythons do eggs, although the maximum numbers, 64 for the boa constrictor and 82 for the anaconda, are slightly lower.[11] As seems to be the universal custom with snakes, the female nearly always ignores her newborn young; however, a female Bahaman boa ate 2 of her litter of 51.[12] A 15-foot anaconda, which produced 19 large abortive eggs and 4 living young, after the birth process was over began to eat the eggs and birth membranes, eventually consuming nearly all of the debris. This is similar to the widespread mammalian habit of eating the afterbirth. Such behavior has not been reported among other live-bearing snakes, although it occurs with lizards. It probably recycles nutrients that would otherwise be lost and also helps eliminate odors that might attract a predator to the exhausted female and her litter.[13]

Reproductive cycles of giant snakes in nature are not well known. Indian pythons in the northern part of their range are reported to mate during the winter months while in semihibernation;[14] however, we suspect that this report may be based on inadequate observation and that most of the mating probably occurs at the end of this period, in late February or early March. Jerry Anderson found pythons brooding eggs in the lower Indus valley in June and July; hatching occurs in this area during late July and August. Boa constrictors in northwestern Mexico have been found mating early in April; the newborn young appear in August and early September.[15] This cycle is common in many temperate-zone snakes. In the equatorial zone, the meager evidence fails to indicate any clear-cut breeding season. Reticulated pythons with eggs have been found in Java in October[16] and African pythons with eggs and hatchlings in Rhodesia in late December and in Angola in late January.[17] Anacondas in Trinidad mate in December and January, and young are born in July and August, but one collected in British Guiana in October gave birth in early January.[18]

The Indian and African pythons are closely enough related that they may occasionally hybridize in captivity. Andrew Koukoulis of Ross Allen's Reptile Institute sent us photographs of the mating of a male African python with a female Indian python. Unfortunately the female died a few weeks later of pneumonia. Her body contained apparently normal eggs.

Food, Feeding, and Growth

Snakes tend to be specialized feeders. Not only are they entirely carnivorous, but many are also highly restricted in their choice of prey. A whole group of snakes in tropical America and a more or less unrelated group in southeastern Asia feed on slugs and snails, and their whole head structure is modified for holding slippery, muscular prey. The mud snake of the southeastern United

States feeds entirely upon two kinds of eel-like salamanders; the snake is found only where these salamanders are plentiful. A widely distributed African snake, when adult, feeds exclusively on bird eggs. Snakes have a decided preference for live food, but many are occasional scavengers. We have seen snakes on the highway eating toads and lizards killed by traffic, and there are reports of snakes' eating animals that were decomposing and alive with maggots.

Although the food of giant snakes in the wild has not been as carefully studied as that of several small species, a good deal of information is on record. It is somewhat biased, however, for observers tend to report the food of adult snakes rather than that of the more numerous but less conspicuous young and also to record large meals rather than those of smaller and probably more typical size. In the first place, a snake with a big meal in its stomach is temporarily handicapped in its movements and more likely to be found and killed. In the second place, if a man kills a python with a conspicuous bulge in its stomach, his curiosity will probably lead him to cut it open, when he will discover a comparatively large animal, say a young pig. A python of the same size with three or four rats in its stomach might not be examined. Unusual or formidable prey is also more likely to be reported. A single report of an adult leopard found in the stomach of an Indian python has been repeated in so many books on snakes that the hurried or uncritical reader might get the impression that leopards are a major food item for these snakes. Nevertheless, some probably valid generalizations can be made about the food preferences of the giant snakes. None seems to be highly selective in its choice of food, and the selection that is made seems to reflect availability more than any other single factor. There are some qualifications, however. Although the African, Indian, and reticulated pythons all have strong aquatic tendencies, there is little evidence that they feed on fish to any appreciable extent. One instance of fish-eating by the more aquatic anaconda is given by the American zoologist and explorer William Beebe in his popular essay, "A Chain of Jungle Life," in which owl kills snake which has eaten fish which has eaten frog.[19] In a more technical account, Beebe reported twenty-seven fishes in anacondas' stomachs.[20] Considering the fondness that many smaller snakes have for frogs and toads, it is surprising that these have rarely been recorded as food for boas and pythons, even very young ones. Another item conspicuously infrequent in the diet of boas and pythons is snakes of other species. Since many species of snakes in other families specialize in eating their kind, it is surprising that only one python, the black-headed python of Australia, is known to eat snakes with any regularity.[21] Lizards, on the other hand, are a favorite food of the Indian python, the boa constrictor, and some of the moderate-sized Australian pythons. Our naturalist friend Chugtai told us that when he was employed by the Karachi Zoo he hung a bag containing two recently purchased large monitor lizards in the pythons' enclosure. A few hours later the bag could not be found, and his superior accused him of selling the lizards for their skins. The affair was cleared up several days later when the largest python was seen in some distress attempting to expell a bulky object from its anus. When recovered, this proved to be the bag that had held the missing lizards.

The eating of a caiman by a captive anaconda was mentioned earlier; caimans have also been found in the stomachs of anacondas taken in the wild.[22] The Australian water python is reported to feed to a considerable extent on young crocodiles.[23]

By far the greatest number of food items eaten by the bigger pythons and boas are mammals and birds. Rodents about the size of house rats are a major food for snakes in the 5- to 10-foot range, but other rodents, ranging in size from mice to the 150-pound capybara, may also be eaten. Porcupines' quills do not always guarantee them immunity; they have been found in the stomachs of the Indian python, African python, and boa constrictor. In parts of

Australia and southeastern Asia, pythons are sometimes kept in storehouses, barns, and boats as a cheap and effective control on rodents. Snakes of 10 feet or more eat small hoofed animals or the young of larger species. In Africa, the many species of antelopes provide such a food source for the python, while in southern Asia there are wild pigs and several species of deer. Domestic pigs, goats, and even calves may be eaten. The largest reliably recorded meal taken by a wild python is a 130-pound impala consumed by an African python measuring 16 feet. A 125-pound pig was found in the stomach of a reticulated python 24 feet long.[24] An anaconda 25 feet 8 inches long had swallowed a pig (probably a peccary) weighing about 100 pounds.[25] Since all these species are reported to grow larger than the individuals figuring in these accounts, the largest snakes can presumably eat animals weighing about 150 pounds. The horns of deer and antelope present quite an obstacle to the snake and occasionally perforate its body after the prey has been swallowed. This mishap is sometimes fatal.

Except for domestic cats and dogs, carnivorous mammals have not often been reported as prey of big snakes. Aside from the often-cited report of the leopard eaten by an Indian python, a full-grown ocelot fell victim to a large boa constrictor, and this snake also eats the mongoose which has been introduced into parts of its range. African pythons are said to eat jackals frequently.[26] Since marsupials were almost the only mammals in the Australia–New Guinea region before the coming of man, it is not surprising that they are an important food for the large snakes of that area. There are reports of the amethystine python's eating kangaroos and wallabies weighing up to 50 pounds.[27] The diamond python and carpet python now feed to a large extent on rats, but they undoubtedly also eat phalangers, bandicoots, and other moderate-sized marsupials. Boa constrictors often eat opossums. The Cuban boa's reputed appetite for bats is unique among the bigger snakes.[28] There are reports of African pythons' eating baboons and reticulated and Indian pythons' taking monkeys, although apparently not frequently. More information on primate-eating habits of giant snakes would be valuable, for it might have some bearing on their tendencies to attack man. Most of the birds eaten by boas and pythons are medium to large ground-nesting species and waterfowl. The main factor in selection seems to be availability. The snakes often prey on domestic chickens and ducks but seldom eat their eggs. Once when we visited a saint's tomb near the East Nara in the Indus valley, we were told that a python had just made off with one of the peacocks sacred to the shrine.

Many boas and pythons have heat-sensing clefts or pits in their lip scales which help them to locate warm-blooded prey. These are particularly conspicuous in some of the moderate-sized arboreal species, such as the New Guinea green python and the emerald tree boa of South America. These structures are analogous to the heat-sensing pit located between the eye and nostril of many vipers. Experiments have shown that the boas' and pythons' pits are about one-quarter as sensitive to temperature differences as those of vipers and are not as efficient as direction finders.[29] Nearly everyone who has had much experience with boas and pythons assumes that the snakes capture most of their prey from ambush. Although these reptiles may seem extremely sluggish, their strike is faster than that of a rattlesnake. Incidentally, a snake's strike is much slower than many people believe. Several times I have heard that if you shoot a rattlesnake, you will always hit it in the head because the snake strikes the bullet in midair. However, the fastest snake strikes recorded are about as rapid as a boxer's jab—approximately 20 feet per second.[30]

Unless the prey is a helpless nestling bird or mammal, all the giant snakes use constriction to kill it. This trait is widespread among snakes, and its high development in the primitive boas and pythons indicates that it must have evolved early in serpent history. Perhaps it developed from the crocodilian and lizard trait of grasping large prey in the jaws and twisting or whirling with it in order to dismember it. If a proto-snake tried this, it would tend to twist its own body around that of the victim, and muscles ordinarily used for locomotion would now apply pressure. This was probably the first technique used by snakes to overcome large prey, and it remains highly successful. On the average, constrictors kill their prey faster than do poisonous snakes and probably have fewer animals escape from them. Several animals may be seized in rapid sequence and constricted at the same time. An African python was seen to catch and kill three jackals in this way.[31] The principal cause of death is suffocation, although compression of the heart probably plays a part as well. The pressure often bursts blood vessels and causes hemorrhage from the nose and mouth, but the victim's bones are seldom broken. It is not "crushed into a shapeless mass," as some exaggerated accounts say, but the snake often does considerable pulling and squeezing of the dead animal in order to mold it into a convenient shape for swallowing.

The growth of large boas and pythons follows a familiar reptilian pattern, being rapid in the period of life up to sexual maturity and then slowing down. Theoretically, growth can continue throughout a reptile's life, but in actuality it stops within a comparatively short time after sexual maturity. The maximum size is genetically determined, but reaching it requires that the reptile have a maximal food intake from early infancy through the early part of its adult life. If we were to try to produce a record boa or python, we would get a brood from the biggest female we could find, fathered by the biggest male, keep the young at about 85° F., feed them all they would eat, and protect them from infection and injury. There are still some significant unknown factors. For example, is a diversified diet better than a uniform one? Are there certain periods during juvenile life when growth is faster than others?

One of the most careful records of growth of a giant snake is that of Clifford Pope's Indian python, "Sylvia," which he received in December 1945. She had been caught in October as an infant in the mess hall of an army camp in Burma. On February 10, 1946, she measured 3 feet 5.5 inches. About a year later she was 6 feet long and weighed 6 pounds 6 ounces. At the end of September 1947, when she was about 2 years old, she was 8 feet 4.5 inches long and weighed 20 pounds 13 ounces. At this time she was probably close to sexual maturity. By the end of May 1948, she had passed the 10-foot mark, but then her growth began to slow down. In October 1956, at an age of about 11 years, she was just short of 12 feet and weighed 56 pounds 8 ounces. Her last recorded measurements, in January 1960, were a length of 13 feet and a weight of 70 pounds.[32]

There is a good deal of published information on the growth of boa constrictors in captivity, and, considering their popularity as pets, there must be a great deal more that is not published. A male grew from a birth length of 18 inches to 6 feet 7 inches in two years, and a female of the same birth length reached 8 feet 10 inches during the same period.[33] A litter of ten young boas were all approximately 21 inches at birth and weighed about 2.5 ounces; ten months later the biggest of the group was 40 inches long and weighed about 1 pound 12 ounces.[34] A female we acquired when she was about 3 feet long has grown to about 6 feet in 4.5 years. Since her diet has depended on a highly uncertain supply of mice, rats, and guinea pigs, she is undoubtedly below optimum size, although a handsome and healthy-looking snake.

Eddie M. Ashmore of Louisville, Kentucky, has provided us with data on three giant snakes in his collection. A reticulated python that measured 6 feet and weighed 2 pounds 12 ounces on September 2, 1969, reached a length of 11 feet 10 inches and a weight of 36.5 pounds by January 1972. During this time it ate slightly over 113 pounds of food, mostly in the form of chickens and rats. A Burmese python measuring 2.5 feet and weighing 14.5

ounces in late November of 1969 increased to 8 feet 11 inches and 20 pounds 1 ounce by January 1972. It accomplished this on an intake of 56 pounds of mice, rats, and chickens. A 2-foot African python weighing 9.5 ounces grew to 7 feet 4 inches and a weight of 12 pounds over an 18-month period, on a food intake of 29.5 pounds.

The Pittsburgh Zoo's "Colossus," a female, measured 22 feet when acquired by the zoo in 1949; in 1957, she measured 28.5 feet and weighed 320 pounds. The San Diego Zoo's "Blue Boy," a big male reticulated python, grew from 14 to 18 feet in slightly less than 18 years, while his weight increased from 103 to 180 pounds. During this time he ate 1638 pounds of food.[35] The difference in size and growth rate between these two pythons may be chiefly one of sexual dimorphism, males never attaining the size and weight reached by females.

Attacks on Man

About 1930 a Burmese jewelry salesman named Maung Chit Chine, while on a shooting trip in the Thaton District near Rangoon, attained immortality of a sort by becoming one of the few human beings to be eaten by a snake. The reptile, described as being between 20 and 30 feet long, must have been a reticulated python. One detail of the story rings false—the man was said to have been swallowed feet first.[36] Although small prey and elongate, flexible prey such as other snakes may be swallowed in almost any position, a snake eating a large and bulky bird or mammal invariably swallows it head first. Of course, a snake dealing with so unfamiliar an object as an adult man might possibly have become confused. Dr. Felix Kopstein, a Dutch physician and herpetologist, in a paper published in 1927, reported that a boy 14 years old was swallowed by a python measuring a bit under 17 feet. This occurred on Sałebabu Island in Indonesia. He also mentions a less well authenticated case of a woman eaten by a python more than 30 feet long. The Dutch herpetologist L. Coomons de Ruiter cites two other instances of Indonesian women eaten by pythons while working in the fields.[37] We have a clipping of a Reuters dispatch datelined Djakarta, April 6, 1964, that tells of a 10-year-old boy eaten by a python at Pakuannratu, Sumatra. At Ross Allen's Reptile Institute we saw a photograph of a reticulated python apparently 15 to 20 feet long with a large bulge in its midsection. An English caption says that the snake had swallowed a girl. The source of the picture is unidentified, but some men in the background seem to be wearing Japanese military uniforms of the World War II period. If even half these accounts are true, the reticulated python is in rare instances a man-eater. It is certainly big enough to swallow a human, and it lives in a part of the world where man is probably the most plentiful large mammal. It may have more of a "primate-eating psychology" than the other giant snakes, but this is pure speculation.

Of the other snakes that grow large enough to be man-eaters, or at least child-eaters, there is a certain amount of evidence indicting the Indian python, despite its usually docile nature. A newspaper story dated April 29, 1969, tells of an 8-year-old boy's being attacked and swallowed by a "huge" python as he was gathering grass in a field near his village in East Pakistan. Unfortunately, the name of the village was not mentioned. The Indian python occurs over most of East Pakistan, but possibly the reticulated python also is found in the extreme southern tip of the province. The late Colonel Frank Wall, an English physician and authority on the snakes of India and Ceylon, tells of a Chinese baby's being eaten by a large snake on Stone Cutters' Island in Hong Kong harbor. His informant was "a young European" who witnessed the incident as a boy, probably about 1900.[38] The Burmese subspecies of the Indian python is the only snake in that area large enough to qualify as the culprit.

There is no well-authenticated account of man-eating by the African python, and this is a bit remarkable, for this species seems to attack man more frequently than does any other giant snake. Walter Rose,[39] C. J. P. Ionides,[40] R. M. Isemonger,[41] and Arthur Loveridge,[42] all reliable herpetologists with much field experience in Africa, have recorded what they believe to be authentic instances of pythons making unprovoked attacks on man, and some other probably equally valid instances have been reported by sportsmen, missionaries, naturalists, and others. Usually the snake seizes the person in its jaws suddenly from a place of concealment. The victim is more often a woman or child than an adult man. The snake then attempts to constrict but breaks off the attack after what often seems to be only token resistance. One attack reported by Loveridge in 1931 ended fatally, the victim being an adult woman in rather poor health. She was evidently washing clothes at the water's edge when seized by the python which was 4.5 meters (about 14 feet 9 inches) long. The snake did not try to swallow her and probably could not have done so. Loveridge was told of another fatal attack where the victim was a nearly grown boy. Both occurred on Ukerewe Island in Lake Victoria, where the local people believe that killing a python will bring dire misfortune. A skin diver photographing animals in Kenya's Mzima Spring was the target of a recent python attack but escaped with nothing worse than a bitten hand.[43]

The immense size of the anaconda would seem to point to its being a potential man-eater, but the record does not bear this out. Henry W. Bates, the English naturalist and explorer who traveled the Amazon in the mid-nineteenth century, told about an attack on a boy whose father left him to watch a canoe beached on a sandbar; he was seized by an anaconda as he was playing in the water but was rescued before he was badly hurt.[44] Another early account from British Guiana tells of an anaconda that attacked an Indian duck hunter and his wife. After wounding the snake with a knife, both escaped.[45] Pope mentions three instances of anacondas' attacking human beings with lethal results. In two of these, the man was apparently dragged under water and drowned but not eaten; in the third, a boy was said to have been swallowed by an anaconda and later disgorged.[46] It almost seems as if the anaconda, accustomed to feeding on aquatic and semi-aquatic creatures, does not quite know what to do with a man if it catches one.

The only account of a large snake's attacking a human being in the Australian region is a newspaper story of a family sleeping in the bush with a three-month-old infant. During the night the mother felt her baby being pulled away from her and reaching out felt the body of a snake. She screamed, and her husband awoke and killed the snake; the child suffered only a few cuts about the face.[47] The amethystine python is the snake most likely to have been involved in this episode.

Big snakes do turn on their would-be captors or killers in self-defense, although less often than might be expected. Usually, they can be killed or caught with a minimum of risk, their main efforts being directed toward escape. They can be surprisingly agile when the situation requires it and can disappear into unbelievably small holes or crevices. Most herpetologists experienced in hunting and handling big snakes agree that a boa or python of 10 to 12 feet is as much snake as a man should try to take on single-handed. This was also the feeling of our Jogi snake-charmer friends in West Pakistan, although their actions sometimes belied it. The wild-animal hunter Charles Mayer told, with some embellishments, how one of his assistants was killed by a large reticulated python they were attempting to capture.[48] In 1961 a South African miner tried to catch a large python. The snake encircled him with two coils and threw him to the ground. He freed himself, apparently without serious injury, but he died the next day. An autopsy showed that his spleen and kidneys were ruptured.[49]

Although there is a great deal of individual variation, the Indian python is generally considered the best-tempered of the giant snakes; the anaconda and reticulated python have a reputation for being dangerous and unpredictable; the African python is somewhere in between, and the amethystine python is an unknown quantity. The two Indian pythons we had as captives in Karachi were quite different in temperament. Neelo, the larger, had a characteristically placid disposition and could be pulled and hauled about by anyone, although dogs made her nervous. Nargis was more irritable and unpleasant. Snakes' kidneys are supposed to secrete mostly uric acid, which is a white, pasty material, but Indian pythons produce considerable quantities of something that, if not liquid urine, is a fair facsimile thereof, and Nargis would douse you liberally whenever she was picked up. Moreover, she bit. Once when I prodded her with my foot she nailed me squarely across the ankle, and some of her teeth went deeply enough to make it sore for several days.

While many big snakes in captivity pass weeks and months in what seems to be a state of good-natured torpor, there is always an element of the unpredictable in their reactions. In March 1961, a Florida showman known as Sailor Katzy was killed by a 21-foot reticulated python. His body was found at the base of a palm tree with the snake coiled in the fronds above. Evidently he was cleaning the snake's cage when it seized him by the wrist and applied lethal constriction before he could call for help. He either staggered into the open compound or was dragged there by the snake. There was no indication that the snake tried to swallow him.[50] Captain Pitman tells of another showman who was killed by a 17-foot African python that he had allowed to coil round him. His bones were said to have been broken in eighty-four places—a highly fanciful touch to an otherwise plausible story.[51] The late C. B. Perkins, former curator of reptiles at the San Diego Zoo, told me that he was once attacked by a large anaconda. The snake caught his arm, but he slammed the cage door on its neck before it could use its constricting coils.

Natural Enemies

Snakes are fragile animals that lack body armor, have a long and vulnerable spinal cord, and no special protection for their brain and vital internal organs. It is not surprising that they have a long list of enemies ranging from waterbugs to tigers. About fifty species of animals have been reliably reported to kill rattlesnakes, despite their being venomous and presumably better defended than nonpoisonous snakes.[52] While special ways of life protect snakes from some enemies and expose them to others, most of their foes are found among the smaller carnivorous mammals, birds of prey, and snakes of other kinds.

When newly hatched or born at lengths of 18 to 38 inches, the giant boas and pythons have about the same natural enemies as other snakes of this size, although, being mostly nocturnal, they are less likely to be preyed upon by hawks and other diurnal birds. As they grow, they soon become too large for many common snake predators. At a length of 10 feet they are essentially immune to attacks from birds and snake-eating snakes, except perhaps the very largest king cobras. The larger crocodilians remain to be reckoned with. Nile crocodiles have been reported to eat adult African pythons,[53] and there is no reason to think that the other big crocodilians do not occasionally feed on giant snakes in Asia and South America. The bigger cats also remain as important potential enemies, although reliable accounts of predation are hard to find. The tiger is known to prey on the Indian python, and the leopard on both African and Asian pythons of moderate size; a really big python, however, is more than a leopard can ordinarily handle. Two python-leopard combats staged a number of years ago for a jungle movie sequence resulted in rapid victories for the snake.[54] Despite the extensive observations on African wild animals, we can find no instance of a lion's feeding on a python. In South America the jaguar is said to do a great deal of its hunting along water courses and thus would be in a position to prey upon all but the largest anacondas, as well as on boa constrictors. To what extent it does so is unknown. The other big American cat, the panther, or mountain lion, is apparently not much of a snake-eater, although it occurs with the boa constrictor in many parts of tropical America. The smaller ocelot is known to be a snake predator and to include the boa constrictor in its diet. The rare and beautiful shadow leopard of southeastern Asia probably takes an occasional python. Outside the cat tribe, the hyena has been reported to feed on the African python, and an Indian python 17.5 feet long was said to have been killed by a pair of otters.[55] Other small carnivores, such as skunks, mongooses, badgers, weasels, and foxes, which occasionally eat considerable numbers of smaller snakes, may be important enemies of small boas and pythons but are no match for the big ones. In fact, the bigger constricting snakes may provide a natural check on mongooses, preventing them from becoming the pests they are on some islands where they have been introduced.

Members of the pig family frequently kill snakes, usually eating them as well. In parts of the United States, domestic pigs have been used to clear areas of rattlesnakes and other venomous species. Among wild pigs, there is an account of a group of peccaries killing a large snake (an anaconda or large boa constrictor), and one of a warthog killing a 12-foot African python.[56] Other hoofed animals, such as deer, antelope, and domestic sheep and goats, have been reported as killing rattlesnakes by jumping on them repeatedly with such speed that the snake cannot retaliate.[57] The motive for these attacks is not clear nor is it known whether they are also directed against nonvenomous snakes.

From Africa come several accounts of battles between large pythons and troops of baboons. These occur when the snake seizes a member of the band with the intent of making a meal of it. Sometimes the python gets a well-earned dinner, but more frequently it is driven off or killed.[58] It is widely believed that monkeys and apes have a profound instinctive fear of snakes of any size. Much of the evidence for this rests on anecdotes and poorly controlled experimental encounters. There is a well-known case of a huge gorilla that escaped but was frightened back into its cage by a quick-thinking keeper who confronted it with a snake in his hands. More carefully controlled experiments with captive apes and monkeys have given equivocal results. One team of experimenters found that young chimpanzees with a parent or trusted keeper showed no fear, while adults were disturbed and reacted with a mixture of aggressiveness and caution. This behavior was thought to be the result of experience and social conditioning, rather than being instinctive.[59] In another experimental study, using wild and captive-reared rhesus monkeys, the monkeys were presented with a toy cat, a toy snake, a live mouse, and a live garter snake. They showed no specific fear response to either the live or the toy snake, although unfamiliar objects, particularly moving ones, aroused apprehension and curiosity in some of the subjects. One monkey treated the test snake so roughly that the experimenters concluded that snakes quickly develop a fear of monkeys.[60] A third group of experimenters found their monkeys terrified of a 14-inch toy snake. By putting the snake in a box that the monkeys were accustomed to open for their food, the experimenters created severely neurotic behavior after thirteen to fifteen experiences. The monkeys either refused to feed or bolted their food, grasped at imaginary objects, and developed organic dysfunctions and disturbed social relationships that interfered seriously with their health and well being.[61]

13. Other Large Snakes

Some 30 to 40 million years ago, the primitive boas and pythons gave rise to a new line of snakes, which in turn developed subdivisions that now dominate the world's snake fauna. The differences separating these groups are internal and are meaningful chiefly to anatomists; however they seem to reflect a rather profound dichotomy. Recent studies that Sylvia Kendall Salanitro and I have done on serum proteins of snakes indicate that the Indian python and the boa constrictor show little serological affinity with the more modern snakes. French researchers studying the boas and more modern snakes of Madagascar have come to similar conclusions.[1] The differences are also reflected in the chemistry of the bile salts and the morphology of the visual cells and chromosomes, as well as in more conventional anatomical features.[2]

A most important innovation of the modern snakes has been the modification of the teeth and salivary glands of some groups into a venom-injecting apparatus. The practical implications of this development have been rather unfortunate for snake classification, since they have led to an exaggerated emphasis on the importance of fang types and a tendency to lump together many kinds of snakes simply because they do not have specialized venom-injecting teeth and are from the human viewpoint harmless. Such snakes need not be closely related in other ways. The serological work that Dr. Salanitro and I have done indicates that the rat snakes and garter snakes of the United States are no more closely related to each other than they are to rattlesnakes and cobras.

The families of modern snakes have not produced any truly giant species, although several are large enough to be impressive. Since the only boas in the United States are two dwarf species, all the large snakes here belong in other groups. The length record seems to be held by a 9-foot bull snake found killed on a highway in Brown County, Texas, in 1949,[3] although some Florida snake hunters claim that the indigo snake reaches an even greater length. The eastern diamondback rattler, which may have reached a length of 8 feet in times past, is the heaviest of North American snakes, with a 7-foot specimen weighing about 18 pounds.[4]

In tropical America, relatives of the U. S. indigo snake, which are known in various places as *cribo, culebra arroyera,* or *sumbadora,* reach a large size. Dr. William E. Duellman, an American herpetologist best known for his work on tree frogs, caught a 9-foot 8-inch specimen in the El Peten region of Guatemala as it was battling with a 5.5 foot boa constrictor. In the sumbadora's stomach was a 37-inch jumping viper that it had apparently had for an appetizer before tackling the boa. In Duellman's words, "the snake-eating capabilities of *Drymarchon* [the sumbadora] can hardly be overestimated."[5]

Dhamans and Their Kin

In the Old World the largest non-venomous snakes of this group are the dhaman, a common Indian snake whose name means "rope" in Sanskrit, and the closely related keeled rat snake of southeastern Asia. The keeled rat snake, which is apparently the larger of the two, is reported to reach a length of over 12 feet, although no specific record supporting this has been given. It is a jungle snake often found in trees. Robert F. Inger, an American herpetologist and ichthyologist, encountered a large one in Borneo as it was stretched out on a limb. He fired at it with a .22-caliber pistol loaded with dust shot, the favorite ammunition of reptile collectors. As he did, "the forest seemed to explode as the gigantic snake thrashed about in a cloud of twigs and leaves before making off, evidently uninjured."[6]

A dhaman was the first snake we caught in Pakistan, and we became fairly well acquainted with the species during our stay in that country. Although our biggest specimen measured a little less than 7 feet, many bigger ones have been reported. The late Frank Wall recorded a specimen just under 10 feet from Myitkyina, Burma, and several over 10 feet from Ceylon, the largest measuring 11 feet 9 inches.[7] However, he did not examine any of these specimens personally.

In arid West Pakistan dhamans are seldom found far from water, and more than one escaped us by swimming off among the lily pads and lotus leaves. Further to the east they live in a greater variety of habitats and often quite close to a dense human population. I was riding a bus through a residential section at the edge of Hong Kong when I saw a fair-sized snake starting across the street. I yanked the signal cord and leaped off the bus with a speed that must have startled some of my fellow passengers. By that time, the snake had already been hit by another vehicle and was thrashing its life away on the pavement. It was a dhaman about 4 feet long.

Dhamans eat almost any animal small enough to be overpowered and are considered of some importance in control of destructive rodents. Those we had in captivity preferred frogs to all other food. Dhamans are not constrictors but subdue active prey by seizing it in their jaws and using a coil of their body to press it against the ground or some solid object until it is dead. This technique is used by some other slender, active snakes, such as the racers of North America.

Dhamans are similar to Asian cobras in size and build and live in much the same sort of country, so it is not surprising that the two are often confused. This is also the case with the keeled rat snake and the king cobra. Dr. Edward H. Taylor, one of the deans of American herpetology, wrote that while he was collecting under police escort in Thailand:

> The captain of police . . . repeatedly warned me of the King Cobra. I suggested that rat snakes were often mistaken for the cobra, but he assured me that he could distinguish the cobra. Later in the day he spotted a large snake, with a length of approximately eight feet, its head reared at the edge of thick jungle. "Cobra," he said. "Now let me see you capture it."
>
> The chance of getting the snake was small. I . . . approached the snake with extreme slowness until I stood within reach of it and with a sudden thrust grabbed it by the head and neck. Despite all my protestations that the snake was a rat snake it remained a King Cobra as far as he was concerned. The story was still extant two years later of my capture of a King Cobra. . . . It makes a better story so.[8]

In parts of India and Ceylon, dhamans and cobras are said to hybridize, or, alternatively, the dhaman is believed to be either the male or the female of the cobra. In point of fact, any interest a king cobra or Asian cobra would show in a dhaman would be gustatory rather than sexual, for both frequently eat snakes.

Venomous Giants

A combination of great size with deadly venom is one that has been achieved by only four of the four hundred or so species of vipers and elapids that make up the major families of venomous snakes. The four are the king cobra, black mamba, taipan, and bushmaster. Although a giant poisonous snake sounds like, and to some extent is, a terrifying animal, it has not proved a highly successful one. For killing prey of small to moderate size—say, from a mouse to a cottontail rabbit—venom seems to be highly efficient. For larger prey, for reasons that are not very evident, constriction seems to have the edge. Moreover, venom seems to confer no great advantage as a defensive weapon in the larger snakes, although there is little direct information on this topic.

Largest and best known of the giant venomous snakes is the king cobra, also known in some of the older literature as the hamadryad. This snake has often been included in the genus *Naja* with the smaller cobras of Asia and Africa, but the presence of large occipital shields (an unusual feature in snakes of any kind), as well as other anatomical and biological peculiarities, justify its being placed in a genus of its own. It occurs over a very wide area from peninsular India to the Himalayan foothills and eastward across the southern edge of China and the regions to the south. Off the Asian mainland, it inhabits Sumatra, Java, Borneo, a few of the other larger islands of Indonesia, and several islands of the Philippines. It reaches its greatest size and abundance in parts of Burma, Thailand, and Malaya; elsewhere it seems to be rare. The maximum size of the king cobra is given as 18 feet 4 inches; this is based on a specimen shot in the Nakon Sritamarat Mountains of Thailand by a Dyak collector. He had lost his way, it was getting dark, and he measured the snake against his gun, so the measurements were not taken under ideal conditions. The snake's head, which was saved, was 98 millimeters (about 4 inches) from the tip of the snout to the end of the occipital shields and 72 millimeters (about 3 inches) in greatest width. The next largest specimen for which we can find a record is one of 16 feet 4 inches.[9] The usual adult size is 7 to 12 feet, with males averaging larger than females.

Over much of its range the king cobra prefers hilly jungle, but in parts of Thailand and Malaya it may live in fairly open country and in cultivated areas. It is a diurnal snake which is largely terrestrial but also climbs and swims well. Early accounts stressed its aggressive nature, and Ditmars considered it "the most dangerous of all living wild creatures."[10] All recent observers have come to almost exactly the opposite view, considering it a rather timid and inoffensive snake. The classic example of its peaceable disposition is an account of the capture of a 15-foot 7-inch specimen on a Singapore golf course. The two men who caught it believed it was a python and seized it by the tail. It made no effective resistance.[11] Characteristically, it rears up in typical cobra fashion, with a quarter to a third of its body off the ground and its narrow hood spread. It has a deep, vibrant hiss that sounds almost like the growl of a dog. It strikes with a comparatively slow, sweeping motion.

Samples of king-cobra venom that we have tested on animals have had about one-third the toxicity of Asian cobra venom, but the average amount per snake is considerably greater in the case of the king cobra. The effects of the venom are much like those of Asian cobra venom, and they can be neutralized quite effectively by some other cobra antivenins as well as by tiger-snake and death-adder antivenins.

Medical literature contains few accounts of bites by king cobras or other giant venomous snakes. There are two schools of thought on why this is so, one maintaining that anyone bitten by one of these formidable monsters will be dead before he can reach medical aid, particularly in the countries where these snakes occur, and the other maintaining that these big snakes, being uncommon and basically timid, rarely bite anyone. There is undoubtedly an element of truth in both views. Dr. H. Alistair Reid, a British physician who has had much clinical experience with snakebite in the Old World tropics, saw one fatal king-cobra bite, inflicted on a Malay farmer who had the misfortune to sit on a 12-foot snake.

Reid also saw a Chinese snake catcher who had been bitten on the finger by a 16-foot king cobra. This man had no generalized poisoning, and the wound on his finger healed in two weeks.[12] Two other recent king-cobra bites have occurred at famous serpentaria where the victims were able to get prompt and skilled medical attention. The more recent took place at the Queen Saovabha Memorial Institute in Bangkok. Although the bitten man was immediately given 130 milliliters (about 4⅓ ounces) of king-cobra antivenin, within 30 minutes the drowsiness, drooping eyelids, impairment in swallowing, and labored breathing characteristic of serious cobra poisoning began. A little over an hour after the bite, his breathing had ceased, he had no tendon reflexes, his pupils no longer reacted to light, and he was to all external appearances a dead man. For 19 hours his breathing was sustained by a mechanical respirator, and he received antivenin, totaling, with the earlier injections, 1150 milliliters, or well over a quart, along with various supportive medications. The bite became infected, and antibiotics were required. He recovered after a month of hospitalization.[13] The other victim was William E. Haast, director of the Miami Serpentarium. Mr. Haast's symptoms, as reported in a nontechnical book, were delayed several hours but were rapid in development and alarming. However, he never completely lost consciousness and was sufficiently recovered to be back at his serpentarium within 24 hours.[14] He had previously been bitten several times by other species of cobras and other elapid snakes and had immunized himself with cobra venom, so he undoubtedly had a higher degree of resistance than the victim in Bangkok.

While the smaller species of cobras eat a great variety of small animals, the king cobra feeds entirely on snakes and monitor lizards. No one has done a study of the food of the king cobra in the wild, but casual observations indicate the snakes eaten are mostly the larger harmless species, such as Asian rat snakes, dhamans, and pythons up to about 10 feet in length. Other cobras and kraits are sometimes eaten, but vipers such as Russell's viper and the Malay pit viper seem to be avoided. This dietary specialization presents a problem to zoos in Europe and the United States, for they must either provide their king cobras with a steady supply of large snakes, which are no longer easily obtainable in densely populated temperate-zone countries, or resort to force-feeding, a procedure hazardous for both snake and keeper.

A unique feature of the king cobra's biology is the female's construction of a nest from vegetable material. James A. Oliver observed and filmed the construction of such a nest by a 13.5 foot specimen in the Bronx Zoo. According to his account:

> When first observed, she was pushing the bamboo litter on the cage floor into a compact pile. She did this by looping the forward section of her body part way around a pile of litter and then drawing her body back in an open loop. She moved the pile into a corner of the cage behind several vertical stalks of bamboo. Soon she had virtually all of the bamboo stalks back in the corner and began pushing sand on the pile by tilting her head . . . and pushing it along like a plow. . . . Moving the bamboo and sand was usually accomplished by movements that resulted in the anterior third or half of the body being employed as a large hook used on its side to drag in the loose material. Occasionally she gripped a pile of leaves with a loop of her body and literally carried it to the growing nest. As material was added the female crawled inside the nest and began crawling round and round in a tight coil, pushing out a chamber from the inside.
>
> She would stop and push a coil out to enlarge or pack down the material on the inner side. Working on and off in this fashion, she finished the nest and laid her eggs in it two days later. She covered the egg clutch with some of the leaves and sticks, and coiled on top of them.[15]

J. H. E. Leakey, an African herpetologist, had the good fortune to make a collecting trip to southern Thailand in May at the height of the king-cobra nesting season. By offering a modest reward ($5.00) to local villagers, he was able to locate sixteen nests. The nests were remarkably similar to the one built by Oliver's captive female, being 2 to 3 feet in diameter and 10 to 20 inches high. All but one were constructed of bamboo litter. They contained 20 to 43 eggs. In every case but one, the female was coiled on or near the mound. Most surprisingly, not one of the females made any attempt to defend her nest or showed any hostility until grasped with the grab-stick or collecting tongs. Moreover, several of the nests were close to footpaths that were used by the natives with full knowledge that the snakes were there. The largest female captured measured 11 feet 8 inches. All the females were thin, as though they had not fed for some time.[16]

Africa has two giant poisonous snakes: the black mamba, which is second in length to the king cobra among venomous snakes, and the Gaboon viper, which, inch for inch, is almost certainly the heaviest. The mambas are a distinctive African group of large, active, whiplike snakes. Although they belong to the elapid family, their fangs are almost like those of vipers, being long and fairly mobile. The head is long and slender, and the fangs are far forward, near the tip of the nose. The three smaller species of mambas are tree snakes, but the black mamba is primarily a ground-dweller. The name of this snake is somewhat of a misnomer, for the animals are never really black, their color when adult varying from olive brown to gunmetal gray. The young are greenish, and this resulted in this species's being confused for many years with the predominantly green tree-dwelling mambas. The maximum length of the black mamba has been stated as 14 feet, although the longest individual for which we can find a definite record is one of 13 feet 7 inches.[17] This is far beyond the average length, which seems to be about 6 to 7 feet for males and 6.5 to 8 feet for females. The largest of the hundreds of black mambas taken by the late C. J. P. Ionides, one of Africa's most skillful herpetological collectors, measured 10 feet 6 inches.[18]

The black mamba inhabits the savanna and semidesert country of south and central Africa. In the east it occurs from southern Sudan and Eritrea to the Cape of Good Hope Province, and in the west from about Walvis Bay in Namibia to about the mouth of the Congo. Quite surprisingly, it has recently turned up in Senegal, some 2500 miles from the main portion of its range.[19]

The black mamba is a diurnal snake that often makes its home in a termite hill or abandoned mammal burrow and may occupy such a site for years. Mambas may well be the fastest of all snakes. Reliable observers have estimated their top speed as 20 miles an hour. While this may be an exaggeration, every herpetologist who has encountered them in the field has commented on their ability to vanish in grass or underbrush like a flash of light. When cornered, the snakes may use this amazing agility to defend themselves. An angry black mamba may assume a cobralike defensive posture with head raised and neck somewhat distended. It opens its mouth, flicks its tongue from side to side in a characteristic manner, and gives an ominous, deep hiss. Its strike is a great deal faster and more accurate than that of a cobra, and it has great reach, up to about 40 percent of its length. The black mamba's venom has about the same toxicity for experimental animals as that of the Asian cobra, and a large mamba probably has no more venom in its glands than a large cobra. However, the mamba may be more dangerous in being a more efficient biter, injecting more venom per bite and injecting it deeper than does the cobra. Black-mamba venom can be neutralized to a considerable extent by cobra or tiger-snake antivenins, but, interestingly enough, venom of the eastern green mamba, which occurs with the black mamba over a considerable part of East Africa, can be neutralized only by a specific mamba antivenin.

Experienced African herpetologists agree that the black mamba, under most circumstances, is an exceptionally timid though nervous snake, but under some conditions it attacks fiercely with little provocation. Ionides tells of catching a mamba more than 10 feet long that had allegedly killed seven people in two years and relates a second-hand account of another snake that accounted for eleven victims. He tells of a man who was fatally bitten by one of two mambas he surprised as they were "fighting." Ionides believes that the snakes were copulating, but they may have been males engaged in the "combat dance" that is known for many species.[20] Dr. David S. Chapman, a South African surgeon, in summarizing his experience with snakebites in Natal and vicinity, recorded seven black mamba bites, all of them fatal, and two other fatal cases in which this snake was the probable culprit.[21] Certainly few species of snakes have a more sinister record. Nonetheless, not all black-mamba bites are fatal. One of Ionides's African assistants, who claimed to have a magic immunity to snake venom, received a one-fang bite and recovered with little trouble.[22] We know of two reptile keepers in the United States who were bitten by captive specimens and survived, one of them without any particularly alarming symptoms. Even more remarkable was the experience of Louis Agassiz Fuertes, a famous animal artist and ornithologist. On his last African expedition he shot a large snake which, because of its narrow head and slender body, he thought was a harmless species. The snake was not quite dead when he picked it up, and it bit him on the finger. He developed no symptoms of poisoning, although the snake, preserved and now in the collection of the Field Museum, was identified as an adult black mamba.[23]

The black mamba's diet consists of birds and mammals, the adult snakes being particularly fond of the curious rabbit-sized animals known as rock dassies, or hyraxes. They also eat rats and may enter houses to hunt for them. Mambas are reported to have an unusually rapid rate of digestion for a snake, and this may be correlated with their generally high degree of activity. Little is known of the black mamba's breeding habits other than that a dozen or so eggs are laid in a termite mound or hollow tree. There is no account of the eggs' being brooded by the female. The newly hatched snakes are 15 to 24 inches long and under optimum conditions may reach a length of 6 feet within a year.[24]

Aged mambas are reported to have difficulty in shedding their head skin, and part of the old slough remains attached to the top and back of the head for a time, forming a sort of plume or crest.[25] It is possible that stories of such snakes, carried by word of mouth from deep in Africa to the Mediterranean seaports, formed part of the basis for the cockatrice and basilisk legends.

While the Gaboon viper, with a maximum length of 6 feet 8.5 inches, falls well under our length limit of 10 feet for a giant snake, it is probably the heaviest of venomous snakes. Large specimens are reported to weigh 7 to 9 kilograms (15.4 to 19.8 pounds), and there would seem to be little doubt that a weight of 20 pounds is occasionally exceeded. The fangs of a big specimen are almost 2 inches long. This big viper lives in forest regions throughout tropical Africa. It is an extraordinarily phlegmatic snake that rarely puts up any resistance when encountered. It has been reported to eat animals as large as monkeys and small antelopes.[26]

The least-known of the giant venomous snakes is the Australian taipan. Even the derivation of its name is uncertain. Some say that it is an aboriginal native name for the reptile; others say that early English settlers applied to the most formidable of Australian snakes the name given to the powerful foreign merchants of China. Although the taipan was formally introduced to science in 1868, for about 50 years it was to some extent confused with two other large species, the mulga snake and the Australian brown snake. Apart from its size, it is not a particularly distinctive-looking snake, being of moderately slender build with a narrow head. It has unusually long and powerful fangs; one of its victims was bitten through a shoe and heavy sock. Average-sized taipans are 6 to 8 feet long, but there is an 11-foot specimen on record from the Cape York Peninsula, and at least a couple of 10-footers have been reported.[27]

The taipan ranges through a good part of Queensland and in the northern part of Australia west at least to Arnhem Land. Although it is said to be common in parts of the Cape York Peninsula and around the shores of the Gulf of Carpentaria, it does not seem to be generally distributed, occurring in pockets of favorable habitat. Another variety of taipan is found in the southern coastal part of Papua, New Guinea.

The taipan is reported to prefer open, grassy country, avoiding both dense forest and desert. It is mostly diurnal and very quick in its movements, usually gliding away in the grass or vanishing down a burrow. When cornered, it throws its body into a series of loops, arches and distends its neck, and flattens its head. It may also lash its tail. It sometimes advances toward its foe, striking quickly and repeatedly.

The venom of the taipan is extremely toxic, being rated by most authorities only slightly below that of the tiger snake, which has the most toxic venom of any land snake known. Since a 6-foot taipan has almost four times as much venom as a 4-foot tiger snake, the chances of recovery following a taipan bite prior to the development of high-potency antivenin were very slim.

Between 1940 and 1960 eight of eleven proved or suspected taipan bites in Queensland were fatal. One of these resulted from an apparently unprovoked attack, the victim being a young woman who was walking from a movie to her home on the outskirts of Cairns about 11 p.m. She felt a large snake seize her ankle and then glide away in the darkness. She ran to a nearby house and summoned aid but died a few hours later with typical symptoms of taipan poisoning. Another victim was bitten on the leg as he was riding a bicycle, and a young Australian herpetologist was killed by a taipan that he had just captured and was trying to stuff into a bag. Most of the other bites took place in the sugar cane fields near Cairns.[28]

Since specific antivenin was developed about 1956, most victims of taipan bites have recovered. If there is anything encouraging about being bitten by a taipan, it is the fact that a period of 1 to 5 or 6 hours elapses between the bite and the onset of life-threatening symptoms, allowing a little time for effective medical treatment to be started. Dr. C. H. Campbell, an Australian physician practicing in Papua, reported six cases of taipan bite, all of which recovered, although one patient was in a mechanical respirator for 10 days.[29] Some of the credit for this good record should go to the efficient air-ambulance service in Australia and Papua, which rapidly transports patients from the remote outback to well-equipped medical centers.

Undoubtedly taipans originally fed on the small to medium-sized marsupials of their homeland, but they readily shifted to the introduced house rat; this now seems to be their principal food in many places and accounts for their frequenting cane fields and the vicinity of municipal garbage dumps. The taipan lays up to 20 relatively large eggs, and the young are about 20 inches long at hatching.[30]

The sole living giant among New World venomous snakes is the bushmaster, although Pleistocene deposits of Florida have yielded fossils of a giant rattlesnake that probably reached a length of 10 to 12 feet.[31] The bushmaster and the rattlesnakes belong to the pit-viper group, which includes about 120 species, most of them native to eastern Asia and the Americas. Interestingly enough, both scientists and laymen have noted a close affinity between the bushmaster and the rattlesnake. A local name for the bushmaster in parts of Central America is *cascabela muda*, or silent rattler, and Carolus Linnaeus, the eighteenth-century Swedish naturalist who founded the modern system of biological nomenclature, put the bushmaster in his genus *Crotalus*, although he stated that it lacked the rattle present in the other four members of the genus known to him.[32] Bushmasters and the larger rattlesnakes have a general similarity in body build, head shape, and large heavily keeled scales. The tip of the bushmaster's tail has a curious burrlike arrangement of scales not seen in any other snake. Contemporary herpetologists hedge regarding the relationship of the bushmaster to the rattlesnakes; most agree that it is close, although the bushmaster is not a direct ancestor of the rattlers.

The maximum length of the bushmaster is given in many books as 12 to 14 feet. We have been unable to find the basis for these records; however, they seem to have originated with German and Dutch naturalists who explored northern South America in the nineteenth century.[33] Ditmars, writing in the early part of the twentieth century, mentions a Costa Rican specimen of 11 feet 4 inches,[34] and a Brazilian specimen 3.6 meters (10 feet 9 inches) is on record.[35] The largest specimens reported in recent years have been between 9 and 10 feet, and the average length is 5 to 7 feet.

Bushmasters are found from southern Nicaragua through the tropical forests of northern South America, along the Brazilian coast almost to Rio de Janeiro, and on the island of Trinidad. For the most part, they inhabit undisturbed jungle, but we have seen specimens collected on banana and cocoa plantations in Costa Rica. They are nocturnal snakes, spending the day in the burrows of armadillos or other animals.

Accounts of the bushmaster's disposition follow a familiar pattern, the early ones stressing its ferocity and later ones its docility, at least under some circumstances. In this connection, Raymond E. Stadelman, a herpetologist with many years' experience in Latin America, wrote of how he obtained a large bushmaster in Colombia:

One of the men was bathing with his wife at a nearby stream when they saw this snake, and he told his wife that such a fine specimen would be very nice for the Stadelman collection. His wife was not enthusiastic, but he insisted that it was a harmless boa. . . . Between the two of them they managed to tie one of his shoestrings around the animal's neck, and started to lead the snake to my house. The shoestring broke, since the snake did not want to go to my house, so the fellow retied the broken lace, and while he led the snake by the string, had his wife shoo and push it along from behind.

In this manner they covered about half the distance to my house, when an Indian . . . somewhat popeyed at what he saw, told them: "Ay compadre, bushmaster very bad snake." The man's wife faded from the picture, and the man himself, with the help of another workman, got a good grip on the snake's neck, and they brought it into the house. They say the poor fellow's knuckles were white and his hand and arm trembling with the force of his efforts by the time he got here. The bushmaster was in worse shape and gasped for air for quite a while before it came around. . . .

I have always had doubts about the famed aggressiveness of the bushmaster. . . . While I consider the fer-de-lance to be very dangerous because it is nervous and apparently bites through fear, I think the bushmaster just doesn't give much of a damn about anything—it certainly does not seem to be afraid of people, but neither does it seem to dislike them.[36]

Another report from Panama is very similar: an American and his wife driving along a road saw a large snake that they believed to be a boa. With their hands and a butterfly net, they got the snake into the trunk of the car and home. The man then transferred the snake to a box and had some difficulty in doing so because the snake persisted in crawling out before he could close the lid. He first slapped the snake's head gently, then grabbed it by the neck and pushed it into the box. When his captive was identified, it turned out to be an adult bushmaster.[37]

As judged by animal experiments, bushmaster venom is one of the least toxic of snake venoms, having only about 1/100 the toxicity of Asian cobra venom. The bushmaster partially compensates for this by having a very large quantity of venom; moreover, the toxicity of a snake venom for mice or other small laboratory animals does not necessarily reflect toxicity for man. The evidence indicates that bushmaster venom differs from other pit-viper venoms sufficiently that it can be only incompletely cross-neutralized by other pit-viper antivenins, and a specific antivenin is generally recommended.

Authentic reports of bushmaster bites in the medical and scientific literature are rare; indeed, one might almost think that these snakes stopped biting people about the time medical journals began printing carefully documented snakebite case histories. There are several accounts in the older literature. Prince Maximilian of Wied-Neuwied told of two Negroes who died from bushmaster bites, one in 6 hours and the other in 12, and said that there were many other instances of fatalities.[38] However, he was writing of the country near Rio de Janeiro, where the bushmaster is very uncommon but another big pit viper, the jararacussu, is relatively plentiful. He said that the bitten men bled from the nose, mouth, and ears; these symptoms are characteristic of bites by the *Bothrops* pit vipers, the group to which the jararacussu belongs. Ditmars mentions that a man bitten in the thigh by an 8-foot bushmaster died in less than 10 minutes but gives no other details.[39] Among 104 snakebite cases admitted to the Hospital San Juan de Dios in San José, Costa Rica, during the middle 1950s, I found none ascribed to the bushmaster.[40] There was one bushmaster bite among 55 snakebite cases treated at the Gorgas Hospital in the Panama Canal Zone between 1925 and 1951; it was not fatal.[41]

Bushmasters feed poorly in captivity, and almost nothing is known of their diet in the wild. Presumably they subsist on small to medium-sized mammals, such as agoutis. The bushmaster lays eggs, a primitive trait among pit vipers, most of which are live-bearing.

14. Snakes as Leather, Food, and Medicine

The beautifully patterned skins that help conceal the giant snakes in their native wilderness have been sought as trophies and for more utilitarian purposes since prehistoric times. Snake skins do not have the durability of crocodilian leather and are not suitable for manufacture into large objects such as suitcases. The popularity of snake skins in the world of Western fashion shows unpredictable fluctuation; it was high in the 1920s and again in the 1960s but now seems to be declining. Nevertheless, the number of skins processed every year is in the millions. Between 1949 and 1957 the United States imported approximately 4 million to 8 million reptile hides annually, Britain an average of 12 million, and other European countries comparably large numbers.[1] Of course many of these were crocodilian and lizard hides, but we estimate that a quarter to a third were snake skins. The proportion has probably risen in recent years, since most crocodilians now receive some degree of protection. The giant snakes whose skins are most frequently used in the leather trade are the African, Indian, and reticulated pythons and to a lesser extent the boa constrictor. Smaller species of boas and pythons are also used, for their skins are tough and the scales have less tendency to curl at the edges. The thick skin of the elephant-trunk or wart snake, a large aquatic species of southeastern Asia, has scales that are juxtaposed rather than overlapping and is widely used in the leather industry under the name of water snake, or karung. Another popular species is the erabu sea snake, which is very plentiful in parts of the Philippine and South China seas. The skins of most other snakes are too small or too thin to make good leather. Cobra-skin articles are often advertised because of the cobra's sinister reputation; they may either be genuine or come from any snake with large, smooth scales. In any case, they do not wear well. Commercial tanning does not preserve the subtle colors of the natural skin, although the pattern may be retained. Nearly all snake skins are bleached and dyed to suit the designer's purpose.

While the idea of eating snakes is repugnant to the great majority of Americans and Europeans, the use of snakes as food is widespread in other parts of the world. Some restaurants in Japan specialize in snake dishes just as restaurants in the United States specialize in sea-food delicacies.[2] All medium-sized to large cities of Taiwan have at least one snake shop selling reptiles for food and medicinal purposes;[3] the same is true of Hong Kong and probably of mainland cities in southern China. A Hong Kong specialty is the dragon-and-tiger banquet, in which the main dishes are snake and civet cat.[4] Since giant snakes are rare or absent in this region, they do not often reach the markets, but Burmese and reticulated pythons are imported and fetch a good price. Poisonous snakes are often eaten. The mamushi, a small and not very dangerous pit viper, is popular in Japan, while the banded krait and Chinese cobra are favorites in Hong Kong markets.

Australian aborigines, who have learned to utilize nearly every food source in their harsh environment, often eat snakes. Some tribes do not eat venomous snakes, but pythons and their eggs are highly prized. Rituals to insure the increase of edible snakes are carried out by several tribes. Women of the Didi cult in the Western Desert lift stones, drop them again, and make snake tracks leading away from them. This is supposed to frighten out spirit snakes which enter mother snakes during the breeding season and are later born.[5] At Obiri in Arnhem Land, tribesmen beat the rock picture of Aniau-tjunu, the water snake, with branches, in order to drive his spirit progeny into nearby water holes.[6]

Some fifteen or so tropical African tribes eat snakes regularly. Because of its large size, the African python is the species most sought after. In the Hindu- and Moslem-dominated parts of Asia there is a general religious proscription against eating snakes, but scattered throughout southern Asia, especially in hilly, jungle areas, are isolated tribes that cling to old animistic faiths. Many of these tribes live by hunting and primitive agriculture and include snakes in their diet. The large species, such as pythons and dhamans, are most often eaten.

Few ethnic groups in tropical America eat snakes. Some of the Guiana Negroes are said to do so, this practice apparently being part of their cultural heritage from Africa. A few Indian tribes living along the great rivers of South America, such as the Mocoretas of the Paraná, will occasionally kill an anaconda or boa constrictor for food.[7]

Rattlesnake meat is sold in specialty food shops throughout the United States, but meat of the big snakes is not, to our knowledge, commercially available in this country. At the International Symposium on Venomous Animals is São Paulo, Brazil, in 1966, we were guests at a luncheon for herpetologists at which boa constrictor was the main course. It was good, although not equal to the royal-snake casserole Madge cooked while we were living in Karachi. We prepared this dish in secrecy and with some trepidation, for our servants would have probably resigned *en masse* had they known we were capable of such an abomination.

The use of snakes in medicine is related to their use as food. It finds its greatest popularity today in the Far East, but viper flesh was used in European medical practice well into the nineteenth century, and rattlesnake oil was a folk remedy in southern Indiana during our childhood. A snake's gall bladder is one of the substances most frequently used in Chinese folk medicine. Usually it is removed immediately from a freshly decapitated snake and swallowed whole with a glass of rice wine. It is said to be a tonic for the heart, eyes, and liver and to relieve the pain of arthritis and rheumatism. The gall bladder of a krait or cobra also counteracts poisons, heals sores, and prevents malnutrition. A mixture of the gall with snake oil is "good for pneumonia, insures virility, improves vision and brightens the eyes." Dried and powdered snake flesh is made into pills that are said to "make stronger muscles and bones, encourage development of new blood, initiate the growth of cells and . . . improve the mental and physical processes of the human body. One can expect to have a spirited and longer life."[8] *Tan-shé*, the python's gall, was a valued Chinese remedy. In Canton, pythons were brought yearly on the fifth day of the fifth month to surgeons who cut out a portion of each snake's "gall" while up to ten men controlled the reptile on a mattress. Tan-shé was considered the only remedy for "mortal infectious fever," and it was so valuable that a portion of each year's supply was paid in tribute to the emperor.[9] In Malaysia people claim that the python has two gall bladders. One of these organs is called the *lampedu idup*, or live gall bladder. They claim that if this organ is cut out and kept, it will develop into a serpent just twice the size of that from what it was taken. The other gall bladder, called the *uler sawah*, is a favorite drug of native doctors. In Selangor and Perak, the "live gall bladder" of a python, when dried, is considered a remedy for smallpox.[10]

In other parts of the world, the fat and oil of snakes is the material most used medicinally. This is doubtless due to the belief that the strength and suppleness of the snake will thus be transferred to the human consumer. Sometimes shed skins, true skins, and vertebrae of snakes are worn as charms against various misfortunes and ailments. All these materials are believed to be more effective if taken from large or highly venomous snakes.

The number of maladies cured or helped by remedies derived from snakes is unknown, but man acquires some peculiar parasites from eating raw or undercooked snake. American physicians who treated Korean soldiers and civilians during the Korean War occasionally saw patients who had painful, itching swellings just under the skin. When these tumors were removed, they contained ribbon-like worms up to a foot long. These were identified as the sparganum or tissue stage of a tapeworm. A bit of medical detective work revealed that all these patients had frequently eaten raw snake meat or organs as a general tonic or as a remedy for such diseases as malaria or typhoid fever. And several kinds of Korean snakes were heavily infected with these same tapeworms.[11] Pentastomids, which have a wormlike appearance but are related to ticks and mites, normally are parasites of reptiles, but they can infect man. Pentastomiasis was found in 45 percent of autopsies on Malayan aborigines. The rate of infection ranged from 100 percent in the Temuans, who regularly eat snake meat, roasted on bamboo spits so that only the outside surface is charred, to zero in the Jakuns, who never eat snakes.[12] Other pentastomid infections have been found among snake-eating peoples of Africa, southeastern Asia, and China. The parasites are usually found in the lungs, liver, and peritoneal cavity. If enough are present, they can cause serious illness.

15. Snake Mythology

Cosmic and Earth Serpents

On January 30, 1972, soldiers guarding the Cambodian capitol, Phnom Penh, lit the sky for a full hour with a fusillade of shells and tracer bullets. They killed two of their comrades and wounded eighty-five others, all in an effort to drive away a malevolent serpent who was swallowing the moon. Commander-in-Chief Lon Nol called the incident a "serious blot on the honor of the Khmer Republic" and threatened to court-martial officers and men responsible for this alarming waste of ammunition. But he was bucking great odds. The moon was eclipsed and the troops' bombardment was intended to save her in a perilous hour. Cambodian people believe that the sun and moon have an evil brother, a cosmic serpent named Reahou. Whenever Reahou attacks either of his siblings, the people make as great a noise as they can, beating on pots, shouting, and striking gongs, in order to warn the heavenly body about to be engulfed and to frighten away the monster.[1]

This custom probably originated in the Nile delta, where, in Egyptian cosmology, Ra, the sun god, had to contend with a giant serpent during part of each 24-hour period. The monster, Apop, personified the spirit of primeval night and the world of death. Each day Ra passed through Apop's realm of darkness, battling for his life so that another day might dawn.[2] In the Papyrus of Ani, Apop is portrayed as a coiled serpent bristling with knives. An accompanying hymn to Ra says: "Thou passeth thru the heights of heaven; thy heart swelleth with joy; the serpent fiend has fallen; his arms are cut off; the knife hath cut asunder his joints."[3]

In Upper Egypt, Ra's priests performed a daily special service to help their deity overcome the dark powers of the serpent. They wrote Apop's name in green ink on new papyrus and wrapped a wax image of the snake in the paper. A priest threw the bundle to the ground and stamped on it with his left foot four times during the day, spitting on the parcel and defiling it. At midnight, the dark midpoint in Ra's journey, the package was put on the sacrificial fire and burned. The priests also made wax images of the foul fiends who helped the serpent. They wrapped the figures in black hair and pierced them with a stone spear. On those days when Apop rose from his dark world to invade Ra's sunlit sky, priests enlisted the help of the populace in making a great noise in order to warn the god and frighten the serpent away.[4]

In 1893 an American anthropologist, Jesse W. Fewkes, reported that the Hopi Indians of the southwestern United States yearly celebrated a katchina festival in honor of the sun during the winter solstice. The ceremonies, called Soyaluna, were designed to prevent the destruction and disappearance of Tawa, the sun, who at this time is threatened by a great plumed serpent. Chiefs quiet and propitiate the hostile deity with prayers to his effigy. Fewkes also noted that in one of the three surviving books of the Maya Indians of Central America, the Maya Codex Cortesianus, a plumed serpent is depicted as swallowing the sun in an eclipse.[5]

On a bluff overlooking Brush Creek in Adams County, Ohio, an earthwork serpent effigy, 1254 feet long, was discovered in the early 1800s and first described in 1848. Later the area was farmed and the mound partially destroyed. In 1886, citizens of Boston contributed $6000 to a private fund for the purchase of the mound and presented it to Harvard University. Frederic W. Putnam of the Peabody Museum, guided by early sketches of the figure, restored it to its original form, and in 1900 Harvard deeded the mound and the surrounding area to the Ohio State Archeological and Historical Society to be preserved as a state memorial. This serpent is ascribed to the Adena Indians, whose mound-building culture flourished in the Mississippi and Ohio valleys from 1000 B.C. to A.D. 700.[6] In its present form the snake's mouth is open and it appears to be swallowing a large egg. Perhaps this is yet another example of the great sky serpent who swallows the sun or the moon.

Sky and earth serpents are a universal theme varying only in details. Naga hill tribes who live in south central India say that the rainbow is the great snake Bhumtaras who rises from his anthill to stop the monsoon. Pardham people in central India see in the rainbow the head of Basuk, the great cobra that supports the world. In Hindu mythology the world serpent Shesha coils about the earth to protect it from malignant influences.[7]

During the Chin Dynasty (221-207 B.C.), the philosopher Fu Huan wrote concerning the white cosmic snake of China: "The potency of the divine serpent is beyond comprehension. It can fly in the sky without wings; it can swim in the sea without fins; it can walk on the earth without feet. It dwells in the air in happiness and self-sufficiency. In due time, it evolves a flaming pearl; whereupon it becomes a dragon."[8]

Samoans claim that originally the sea and sky were stuck together. A giant serpent squeezed into the join and pried them apart. Standing on its tail, it raised the sky from the earth. Gods cut the snake into pieces which fell into the sea and became islands. The snake's blood, splattered on the sky, became the stars.[9]

Palau Islanders tell of a couple who wished to leave the earth. They climbed up the body and neck of a serpent which connected the sea with the moon and there set up housekeeping, being nourished by the fruit of a solitary orange tree.[10]

In the American southwest the Sia Indians boxed the compass

with imaginary giant snakes. A great plumed serpent guarded the north. He and his counterpart in the south influenced clouds, controlled water supply, and induced regeneration of life. The serpents of the east and west exerted steadying influences on the sun and moon. Hu'waka, the great snake of heaven, was said to have a body of crystal, so piercingly brilliant that no man could look at him. Ya'ai, the earth snake, was spotted and splotched, and he hid himself shyly underground. Six snake societies in Sia culture invoked the cardinal snakes with incantations and songs. One song, translated, ran: "Let the people of the white floating clouds, the lightning, thunder, and cloud people, come and work for us that the waters of the six great springs may impregnate our mother the earth, that she may give us the fruits of her being.[11]

In pre-Columbian North America Delaware, Shawnee, and related Indian tribes believed that a giant snake who lived under the earth was the essence and author of evil. This reptile was the guardian spirit of wicked shamans. It was responsible for the destructive deluge described in Delaware mythology, during which the tribes took refuge on the back of a huge turtle. When they arrived at the center of the earth, twelve powerful shamans of the turtle clan overcame the power of the evil serpent by means of magical songs, incantation, the use of sacred objects, and potent medicine.[12]

In historical times, snakes continued to represent trouble to the Delawares, and trouble represented snakes. A fragment of their history from 1600, translated from the Lenapi language, reads:

> Halas, halas, we know now who they are; these Wapsinis [white men] who then came out of the sea to rob us of our land; starving wretches with smiles they came, but soon became snakes. . . . They were well received and fed with corn; but no land was ever sold, we never sold any; they were all allowed to live with us as our friends and allies because they

> were hungry and [were] thought the children of Sunland and not Snakes or children of snakes . . . they killed our chiefs . . . they follow everywhere and we made war on them till they sent Makhiakho [Black Snake—General Anthony Wayne] who made strong war . . . we resolved to exchange our lands and return at last beyond the Masispek [muddy water, or Mississippi] near our ancient Seat. We shall be near our foes the Wakon [Ozages], but they are better than the Yankwiakon [English Snakers].[13]

An ancient American Indian creation symbol, the Tsi-kuli, is currently popular in the United States as a Christmas decoration. It is a diamond-shaped figure formed of yarn in many colors on two pieces of crossed wood. Called the *Ojos de Dios*, "eyes of God," it is inaccurately assumed to be an Indian representation of the Christian cross. Actually, similar figures made of yucca have been excavated in prehistoric cliff dwellings in Arizona and in Inca mummy burials in Peru. Seventeenth-century Spanish explorers found them on Indian graves in northwestern Mexico and thought them evidences of pre-Columbian Christian influences.

The figure of the diamond represents the universe. According to the legend, Tlaloc the Rain God and his people wished to dance, and in their sky and water home there was no suitable place. They asked the mother goddess for help. She caught two arrows from the morning star and crossed them, tying them securely together. Then she wound a great serpent around the joint to stabilize it. Finally she cut her own hair and braided it about the serpent. The delighted gods brought a ball of earth and spread it on this foundation. As they stepped on it to dance, the earth began to spread and widen and land was formed. On this the goddess created all life.[14]

In Europe, the most fearsome mythical serpents are found in Baltic mythology. It seems unlikely that early Scandinavian peoples had any experience with real giant snakes; nevertheless, their legends graphically express man's fear and respect for both large and venomous snakes. Some of these impressive serpents may be found in the Prose Edda, a Norse saga of creation and a prophecy for the world's destruction which reads like a poet's metaphoric blueprint for nuclear disaster.

In the beginning were fire and ice. In Niflheim, realm of ice, a terrible serpent named Nidhoggr lived in a spring. Twelve rivers of venom flowed from his fangs, interrupted only when the snake paused to gnaw on the roots of Yggdrasil, the world tree.[15]

Odin, chief of the gods, worked diligently toward an orderly society, but his efforts were often frustrated by Loki, a disruptive, alienated deity who was intimately connected with the forces of chaos and the infernal world. Loki coupled with a giantess, Angurboda (anguish-boding), who bore him three monstrous children: Fenir, a malevolent wolf, a repellent girl child named Hel, and Jormungar, a giant snake. Odin bound the wolf securely and cast Hel into the underworld where she became the goddess of death, supervising the damned as they bathe and swim in the venomous rivers flowing from Nidhoggr's fangs. He threw Jormungar into the ocean, where the snake grew until he could cast a coil about all the lands known to man and put the end of his tail in his mouth. He is the Midgard Serpent who waits in the sea for man's destruction and the twilight of the gods.[16]

Yggdrasil, the ash tree, fills the universe. Its fresh green leaves overshadow Asgard, home of the gods. Its roots extend through the earth into Niflheim where Nidhoggr and a mass of vipers gnaw at them. Norns, or fates, water the tree every day and sit in its shade to weave the web of life. Eventually the primal serpent will sever the root of the world tree. Then the gods will ride down the rainbow bridge from Asgard to battle demons rising from the infernal regions. The Midgard Serpent will twist its way to land to meet Thor, the thunderer, who will strike it a mortal blow with his great hammer, but Thor too will die, drowned in floods of venom from the dying monster's jaws. Fire will engulf the world tree, and the earth, all life destroyed, will sink forever beneath the boiling water.[17]

Two mythical sky serpents are recognized as constellations by modern astronomers: Hydra, the water snake, and Ophiuchus, the man who holds a great serpent. Hydra is the largest of all the constellations, spanning a quarter of the sky. It lies under the foot of the constellation Leo, the lion. The other figure, called Serpentarius by the Greeks, has been described as representing Apollo's son Asklepios, the divine physician whose healing arts were too successful to please Hades, god of the dead. Hades noted a sharp decline in the population growth of his underworld kingdom and induced the Olympian gods to banish Asklepios to the sky.[18]

The Python of the Sun

In preclassical Greece, Gaea, the earth mother, pronounced oracles through her priestess at a shrine at Delphi on the steep lower slopes of Mount Parnassus.[19] A large snake was kept in the sanctuary to represent the goddess in her serpent or chthonic (underworld) form. Gaea's snake was called Python; whether or not it was a python in the current use of the term is uncertain. African pythons might have reached Greece through trade with Egypt but certainly were not native to Greece. The largest snake native to the area is the Caspian racer, which may reach 7 feet in length. Whatever it was, Gaea's serpent gave its name to a whole group of snakes.

The name Python was derived from the Greek word for "rotting serpent corpse"[20] and may have originated as a descriptive term for the gases which emitted from a rocky cleft in the shrine, inspiring the priestess in her pronouncements. Excavations have shown that the shrine and ancient city of Delphi have been continuously inhabited since the early fourteenth century B.C., and Delphi itself was once known as Pytho.[21]

When Zeus produced Athena from his head and thus demonstrated to his consort, Hera, how unessential she and all women could be, the goddess was offended and enraged. She smote the earth with the palm of her hand and called upon Gaea and the old chthonic deities to grant her a son mightier than Zeus. She bore Typhaon, and Gaea sent Python to nurse the child. The boy matured into a monster with a hundred dragons sprouting from his shoulders and countless vipers from his thighs. He continued to grow larger and more repulsive until the gods took fright and fled to Egypt, leaving the problem to Zeus. Zeus beat down the fiend with thunderbolts and covered his stunned body with the volcano, Aetna. Typhaon's restless energy still triggers eruptions and causes earthquakes.[22]

As the influence of the old earth gods declined and they were replaced by the Olympians, the sun god Apollo appropriated Gaea's shrine on Parnassus. When the Python tried to interfere, Apollo killed it. To appease its divine shade, he founded expiatory games in its honor. These were the Pythic games of Delphi, held every eight years along with a festival to commemorate the Python's death. Apollo decreed that the priestesses who prophesied in his honor should be called Pythia, or Pythoness.[23]

The term python became synonomous with soothsayer or oracle, and many rival shrines sprang up in Greece and the Middle East, featuring giant snakes, actual or in effigy, and promising divine help for human problems.

Apollo's temple, superseding that of Gaea, was built over the cleft which emitted volcanic gases. These vapors probably included hydrogen sulfide, methane, and nitrous oxide. They threw the priestess into convulsions, so that she often had to be forcibly held on the sacred tripod seat by attendants. Some women died while on the tripod, others immediately after leaving it. The Pythoness's incoherent ravings were interpreted by Apollo's priests, a device whereby they exerted tremendous political influence on the Greeks, and through them on the known world—an influence which has been compared to that of the medieval popes in Rome.[24]

Apollo's shrine waxed rich and tempting. One valuable trophy dedicated to the Pythoness was a bronze column of three twisted serpents holding a golden bowl on their heads. It was removed from Delphi in about 330 A.D. by the emperor Constantine and set up in the Hippodrome of Constantinople. The column seems to have incited acute anxiety and rage in Christians and Moslems alike. In the tenth century, a misguided Patriarch, offended by the column, had two of the snakes' heads broken off. In 1453, on the day that Islamic troops took the city, Sultan Mehmed II, riding across the hippodrome toward Saint Sophia, caught sight of the column, and with a shout of "What heathen idol is this?" rose in his stirrups and hurled his iron mace, breaking off the last remaining head.[25]

Living Gods

The gap between mythical and real giant snakes narrows in Africa where living pythons are often still protected, cherished, and worshipped as benevolent gods, ancestral spirits, and phallic symbols. Python worship, which once extended along Africa's west coast from Senegal into Nigeria, furnished self-justification to

European and American slave traders, who believed blacks should be rescued from their idolatry and exposed to an enlightened and merciful faith. Once established in America, many Africans resisted Christian proselytizing and continued to practice the ophiolatry of their homelands. These cults, called Voodoo, were once common in the southern United States and may still be found in some Caribbean islands and in northern Brazil.

Formerly in what is now Liberia a python god named Danhgbi was represented in each major village by a living snake. Priestesses served the god. Their chief business was prostitution, and their chief duties, aside from feeding the python, were to furnish music and entertainment for festivals in the god's honor. Children born in the temple compound were considered to have been fathered by the snake.[26]

In Dahomey, the python was called Dangbe, or "life-giving-snake." As the god of wisdom, it officiated at the birth of the parents of mankind who, like puppies, were born blind. The python touched their eyes with his tongue and gave them sight.

Dahomey pythons were associated with Legba, a phallic deity, and their temples were located in the center of each community. Holes left in the sides of the conical shrine let the snakes, several as a rule, come and go freely. A native who found a straying god prostrated himself before it and covered his head with dust, saying: "You are my master; you are my father; my mother; my head belongs to you; be propitious to me."[27]

After Europeans established trading posts in Dahomey, python priests, for a fee, rescued errant pythons from European houses. A man who killed a python was usually burned alive. The pythons were furnished with many wives who fed and watered them, wove grass mats for them, decorated the temple, and offered sacrifices. Any child who chanced to touch one of the sacred snakes was required to spend a year in a special school where he learned the songs, dances, and rituals of the cult.[28]

William Bosman, a Dutch merchant and government official stationed in Guinea in the late 1600s, described his experiences with an active snake cult of the Fida people in great detail. Bosman wrote:

> . . . they invoke the snake in excessively wet, dry, or barren seasons; on all occasions relating to their government and the preservation of their cattle, in one word, in all necessities and difficulties . . . very great offerings are made to it [the python] especially by the king by instigation of the priests. . . .
>
> The snake-house . . . is situated about two miles from the King's village, and built under a very beautiful lofty tree, in which the chief and largest of all the snakes resides. He is a sort of grandfather to all the rest; is represented as thick as a man, and of an unmeasurable length. He must also be very old, for they report that they found him a great number of years past . . . at which being overjoyed, they welcomed their new-come god with all expressible signs of reverence and high veneration, and carried him upon a silken carpet to the snake-house where he is at present. . . .
>
> The first time that I came to Fida to trade, I was assured that as soon as a girl was touched by the snake, she unavoidably ran mad; though it was but a sort of holy madness. . . .
>
> During my residence at Fida, the King caused his daughter to be seized by the snake . . . to be carried to the snake-house and confined . . . on the day of her delivery, she was brought out in a very splendid manner and carried, with all the other girls which followed her, to the King's court. She was naked except only a silk scarf which was passed betwixt her legs and richly adorned with two sorts of coral. . . . Whilst she was here, she was guilty of all manner of extravagances, during the playing of several musical instruments; which sort of madness remained upon her. . . .

> A long time past, when the English first began to trade here, there happened a very remarkable and tragical event. An English captain being landed, some of his men and part of his cargo, they found a snake in their house which they immediately killed and without the least scruple, threw the dead snake out the door; where being found by the Negros in the morning . . . they furiously fell on the English, killed them all and burned their house and goods. This struck such a terror into that nation that for a long time they refrained coming hither and traded at other places. . . .
>
> In the year 1697 . . . a hog, being bitten by a snake, in revenge seized and devoured him in sight of the Negros who were not near enough to prevent him. Upon this the priests all complained to the King; but the hog could not defend himself and had no advocate and the priests begged of the King to publish a royal order that all the hogs in his kingdom should forthwith be killed . . . the King's command was published all over the country, and it was not a little diverting to see thousands of Blacks armed with swords and clubs to execute the order. . . . The slaughter went on, and nothing was heard but the dismal sound of kill, kill, kill which cost many an honest hog his life. And doubtless the whole race had been utterly extirpated if the King, perhaps moved to it by some lovers of bacon, had not recalled his order.[29]

A belief that humans transmigrate into animals after death is ubiquitous in Africa. The British anthropologist Sir James G. Frazer reports that in a rest house near Fife in Northern Rhodesia a tamed python waxed fat on offerings of fowls and sour beer brought to it by Wiwa natives who thought it housed the spirit of their dead chieftain. Kafir kings were said to become pythons or mambas after death while their queens became "tree iguanas." Bahima people of Ankole in Central Africa thought that their dead princes and princesses became snakes. A special shrine was made available to these spirits where priests fed and cared for all visiting snakes.[30]

In Madagascar the souls of dead Betsileo nobles, after a profoundly unsavory interval, transmigrated into boas. The corpse of the nobleman was strapped to the central pillar of his house and allowed to rot. An earthen pot was placed under the body to catch the products of decomposition. When the *Fanany* (worms) appeared in the bowl, the largest was selected as containing the spirit of the dead man. It was placed in a jug. The body was then buried with the jug sunk into the earth on top the grave with a stick placed in its mouth for the convenience of the worm. The worm was expected gradually to turn into a boa, which would eventually return to visit the nobleman's village. When a boa did appear, the people knelt and saluted it. They spread silk for it to crawl on and sacrificed an ox in the public square, allowing the snake to drink the blood.[31] The British scientist and traveler James Sibree, in his treatise on Madagascar, says that the blood was taken from the ox's ear and mixed with rum. If the boa drank this, there could be no doubt as to its identity.[32]

The Reverend J. Richardson, a missionary, reported:

> I have heard from the lips of the chief prince of one of the tribes when his mother was dead, "She has not yet appeared in the pot and I cannot bury her body." Of this prince's mother I know that for nearly three months from the time of her death . . . until the *"fánàny"* appeared, the people of the whole district were not allowed to dig or plant. There was danger of a famine and the Hova authorities were obliged to interfere and hasten the appearance of the *"fánàny."*[33]

One African python cult was centered near Lake Victoria in east central Africa. There the python god, named Selwanga, was represented by a real python which lived in a temple built for its convenience. A woman served the snake. Her daily duties included filling a wooden bowl with milk and holding it while the python drank. Young couples planning to be married came to the python for its blessing on their union, and barren women often traveled from far places to importune the snake for children.[34]

In South Africa the python's phallic role was expressed in a Basuto ceremony. Girls at their first menses were taken to a river by older women of the village. Each girl was hidden separately in the bends and turns of the stream and told to expect a giant snake to visit her. Afterward they covered their faces with little masks of straw and returned to the village.[35]

In some parts of Africa the natives kill and eat pythons. Uganda natives, however, will not harm a female python brooding her eggs. Such a snake is *lubale*, or sacred, and is of general interest to the community. The people visit her regularly and leave offerings of eggs, cowry shells, tomatoes, roast coffee beans, and money.[36]

In Zululand killing a python carries a special risk. The Kafirs believe that the gall of a man who has killed a python will confer to anyone who swallows it wonderful virility, courage, and longevity. F. W. Fitzsimons, a South African herpetologist, tells about a Natal Kafir named Impunzi who was entrusted with his master's hounds. One day, to save a dog's life, Impunzi killed a large python and unwisely boasted about it to his friends. When the news spread to surrounding villages, one old chief, who had a harem of fifteen wives and had recently added three more comely maids to his household, coveted Impunzi's gall. He called a local female witch doctor and offered her ten fat cows to get it for him. The witch doctor called a meeting of her devotces and told them that their chief was ill and might die because the spirit of the python which Impunzi killed was calling for revenge. She suggested that Impunzi be sacrificed for the general welfare of the community. Village men ambushed Impunzi and cut out his liver.[37]

Pythons from the Dreaming

Nowhere in the world are pythons, real and imaginary, more important to people than in Australia, where blackfellows say that even before the "dreaming," the period during which their tribal myths and religious rites evolved, all creation lay in the belly of Eingana, a rainbow python.[38] After the world was formed, giant pythons guarded the billabongs, traced the courses of rivers, and coupled with some of the early women in order to establish families and clans.

In the creation story of the Liyagalawumirri people, during the dreaming two sisters walked across the continent to Arnhem Land, naming the sun, moon, and stars and all living things as they touched them with their yam-digging sticks. At nightfall they camped near the water hole of Yulungurr, a rainbow python, and bathed in the water. The younger sister was menstruating and her blood polluted the water. This so infuriated the python that he swallowed the sisters and then kept on swallowing until he had ingested every living thing. He rose to the sky where other great pythons chid him for his loss of control. Yulunggur then plunged back to earth and disgorged everything back into its proper place.[39]

Gunwinggu people forbid menstruating girls and women to eat water plants and seeds for fear of enraging the python who guards their water supply. During her menses a woman stays close to a fire far from the water hole and takes care not to attract the snake's attention. When a young girl is near the end of her first menstruation, a python is painted between her breasts. She then swings from the branch of a tree by her hands, while her mother slaps her belly with a pack of cold wet sand wrapped in paperbark, in order to dry up the menstrual flow. If the python should smell her blood he might kill the children in the encampment.[40]

Aborigines south of the Daly River isolate a girl at her first menses in a specially built hut in order to protect her from the python. Afterward her mother burns the hut. When the girl's period is over, village matrons paint her with red ocher and white clay and solemnly escort her back to the village. The girl's mother meets her, crying and gashing her own head with a hatchet.[41]

In East Arnhem Land, tribal elders warn pubescent boys that the great python Julunggul smells their foreskins and is coming to swallow them. The terrified children cling to their mothers, who form a circle around them and make a great show of fighting off the men with spears and clubs. The men forcibly carry off the boys into the bush where they draw a little blood from each initiate's arm to offer to the snake. Bull roarers howl and the great bamboo pipes drone as the men smear the boys with ocher and blood to symbolize the state of ritual death they are about to undergo. Each boy in turn is stretched out on a table formed of the backs of older men, and his foreskin is excised with a flake of stone.

Meanwhile, other men dig a crescent-shaped trench which represents a uterus, decorating its sides with pictures of pythons. Members of the snake totem huddle in the trench and writhe in simulation of birth as the circumcised boys emerge from the bush, being "vomited forth by the python." In West Arnhem Land, similar initiation ceremonies are held to the beat of a long wooden gong called the *urbar*, which represents the penis of the rainbow python.[42]

Human reproduction is wholly the responsibility of the rainbow pythons. Murinbata people say that Kunmanngur, a rainbow snake, has a space cleared on one side of his water hole where he stores the spirits of male children in stones. Female spirits he keeps with him in the water. He may also use fish, goannas, tortoises, and geese to contain spirits. When a hunter returns to camp after a long journey, a spirit child, *Ngarit-ngarit*, enters his wife under the nail of her big toe and she becomes pregnant. Sometimes the python releases a few of the girl spirits to amuse him in the water. If a hunter tries to capture one of these elusive females, the snake will crush him in its coils.[43]

East across the Pacific

On Fiji a mythical giant serpent, Ndgendi, was believed to be the creator of all things. Ndgendi lived in a cave at Raki Raki on the northeast side of Viti Levu. He taught man to make fire by rubbing two sticks together. When his people became impious and depraved, he punished them by sending a devastating flood. Fijians sacrificed to the snake in order to induce him to send rain and bountiful crops. One offering listed 200 pigs and 100 turtles. Men faced with a scanty yam harvest sometimes sacrificed their wives.[44]

Fijians say that a skeptical chieftain once visited the shrine with an offering of fish. When the serpent graciously appeared to accept the fish, the chief treacherously shot at the god with his bow and arrow. Then, horrified at what he had done, he fled, pursued by a terrible voice which cried: "Naught but serpents! Naught but serpents!"

By the time the chief had reached his home he decided that nothing was really amiss, and being hungry, he called for dinner. When the pot was uncovered, it was full of living snakes. Baffled and frightened, he raised his water jar to his lips only to find it full of writhing serpents. He lay down to rest on his bed but the hissing of uncountable reptiles made sleep impossible. Finally he sought out a priest and confessed what he had done. Together they made a generous propitiatory offering to Ndgendi and received pardon.[45]

Ophiolatry, which extended across Asia to Fiji, suffered a sea change in New Zealand and eastern Micronesia. There the roles of the snake are played by a giant eel, but the motifs remain the same. The American anthropologist Joseph Campbell, in his "Great Serpent of the Earliest Planters," suggests that phallic eel myths in the Pacific may extend back as far as 7500 B.C., and the legends were carried to Easter Island by Marquesan emigrants in about the fourth century A.D.[46]

The primal myth describes how Hina, the Polynesian Eve, was bathing and playing in the water when a smooth eel began to caress her body, slipping his sinuous length between her thighs in the "place where pleasure is." When Hina was erotically aroused, the eel transformed himself into a splendid young man so as to cohabit with her. Afterward, however, Hina reconsidered. She decapitated her eel lover and buried his head in the sand. A few days later, a green sprout appeared which quickly grew into a tall coconut palm. When the first ripened fruit fell, Hina husked it and found the two small eyes and mouth of her lover on the brown nut.[47]

Nagas and Jogis

Pythons are usually a sideline with snake charmers in Pakistan and India, the Jogis (snake men) preferring to work with cobras. However, I once asked Maulana, a Jogi friend, why we saw only young and beautiful women in his village. He replied: "When they grow old, Memsahib, we give them a python and send them out to beg."

It must be true, for over a four-year period in Karachi numerous malnourished old women stopped by our compound gate and offered to show us pythons. The snakes were generally in as bad shape as the women, or worse, and we felt sad for both.

Jogi men usually work with cobras, which are traditionally descended from the serpent gods of prehistoric India. These gods, called Nagas, were supernatural cobras of great size, some of which were believed to be as much as 400 miles long. They were said to guard rivers and lakes, control volcanic eruptions, cause earthquakes, usher in the annual monsoons, and harry mankind with plagues and famine. At the same time, with the ambivalence characteristic of ancient deities, they took a personal interest in the conception and birth of human children. Some Nagas assumed human form and mated with women in order to establish princely houses. Shesha, the first and greatest of these serpent gods, coiled around the earth to protect it from malignant influences.[48]

One Naga and his serpent people lived on Chacra-giri, a mountain far to the east. The breath of these snakes swept down the mountain's slopes in a fiery poisonous wind that burned and destroyed animals and plants and left desert the lands which it touched. Two saints, Agasti and Astica, traveled to the mountain and by means of powerful magic subdued the serpents and reduced their size so that they would fit into little clay pots. These saints' spiritual descendants are today's Jogi tribes who catch cobras and display them in little pots and baskets, thus demonstrating the ascendancy of man's good over the snake's evil.[49]

The legendary mountain Chacra-giri may correspond to an extinct volcano, Popa, in central Burma, which is the mythological home of the great serpent who governs king cobras, the world's largest venomous reptiles. Snake charming in Burma is done by women, who work exclusively with king cobras. The performance is highly stylized and suggests that it may be a rite once practiced by priestesses of an ancient ophiolatrous cult. Before performing with her snakes, the woman dresses in special ceremonial clothes and whitens her face with rice powder. She begins each show with a prayer to the serpent king of the volcano Popa, promising to release her snakes unharmed after a given period of time. Her performance demonstrates her ability to anticipate and avoid the snake's relatively slow strike and her sound knowledge of what the reptile can and cannot do. After she has thoroughly confused and frustrated the giant king cobra, she ends her show by kissing the snake three times on its spread hood.[50]

Although Hinduism, Buddhism, and Islam have successively swept across Indian and eastern Asia, the ancient serpent gods still retain some of their old power and influence. Ceremonies in their honor are annually observed in the Punjab area of northwestern India and Pakistan. Each year, in July, people in southern India, in Madras, Kerala, and along the Malabar coast celebrate the Nagapanchami, a great festival in honor of Devi, an earth goddess who once stood guard in the shape of a giant cobra in order to protect men from venomous snakes.[51]

Formerly, and perhaps still today in parts of India, architects, when planning an important structure, called in an astrologer to show what spot in the foundation would be exactly over the head of the great Naga which supports the world. A mason then fashioned a little peg from the wood of the Khadira tree and using a coconut as a hammer, drove the stake into the ground, thus pinning the serpent's head in place and insuring the building's stability. This custom probably dates back to pre-Vedic times, before 1600 B.C. During the Gupta period, A.D. 320–470, large posts, sometimes cast in iron, were used for this purpose when Indian artisans laid the foundations of temples and palaces. The most famous of these is the iron pillar of Delhi set up in A.D. 400 by Kamaragupta I. It is over 23 feet high and is estimated to weigh about 6 tons. Smaller pillars have been found at Dhar in south central India and at Mount Abu.[52] Our Jogi friends in Karachi told us that such a pillar was set in the wall of Thatta, a once-proud city on the Indus river some fifty miles east of Karachi. When the local Mir (ruler) had the post pulled up, the city immediately began to decline and fall into ruins.

Centuries of invasions across the Indian subcontinent have not succeeded in extirpating or even significantly changing the aboriginal Gonds who constitute the Naga hill tribes of south central India. Many of the Gonds' religious and social customs are unique. Social scientists have been especially interested in their youth culture. Young people of both sexes live together in a kind of dormitory, called the *ghotul*, where they are trained by community teachers in civic responsibilities but otherwise are left free to their own social life and premarital sexual relations.

Gonds say that the first ghotul was established by the god Lingo Pen. Its central pillar was a huge python, the beams were dhamans, and the roof poles were *maha-mandal* snakes. The framework of the roof was formed of kraits braided with vipers and covered with tails of peacocks. Seats in the ghotul were crocodiles, and a swing hung from the roof was supported by ropes made of *pirtitti* snakes. The whole structure was decorated with bulbul plumes. A more phallically suggestive motif would be difficult to imagine.[53]

How the Python Lost Its Venom

Man has evidently puzzled as to why the significant and fearsome python which had so much going for it was not venomous too. At least two legends explain why this is not so.

Sea Dyaks of Sarawak claim that the python was originally the most venomous of snakes. One day the python caught a man robbing his fish pond and bit him. However, the man recovered, and the snake, in a burst of temper, vomited all of his poison into the sea. The sea snakes eagerly ate it all up and became universally poisonous themselves.[54]

The second legend, from Burma, tells how the original python, who was perfectly white, was also the first and only venomous snake. The python fell in love with and seduced a woodcutter's wife named Eu, who was much taken with him. After the couple had eloped, Eu wove a marvelously designed and variegated coat for her lover. He was greatly pleased with it, and his descendants wear it to this day.

Their insular life in the jungle soon became tedious, and the python began to experiment with his venom. He could kill a man by biting the human's footprint long after he had passed. There was no way for the python to know whether or not the venom was really working, so he bribed a crow to check results for him. The crow misinterpreted funeral preparations in the victim's village as a festive occasion and told the python that since the people were rejoicing, the poison was clearly ineffective. Disgruntled, the snake climbed a tree and vomited up all his venom. The liquid fell on the tree, which itself became poisonous, and its juice is used to poison arrows today. The venom also fell on the cobra, the water snake, and the frog.

Eu chided her python for spreading death about so thoughtlessly, and finally the snake called together those creatures who had received the poison and asked them not to use their gift carelessly. The cobra agreed and said, "If there be transgression so as to dazzle my eyes, to make my tears fall seven times in one day, I will bite, but only then." But the water snake and the frog laughed at the python and said that they would bite whomever they pleased. The python chased them into the water where the venom was all washed away.[55]

Honorable-Lords-Serpents

In the narrative of Japan's cosmology, snake legends remain as fragments of an ancient ophiolatry practiced by the people long before they migrated to the archipelago. Throughout the centuries, Shinto and later Buddhist priests have recognized the snake's sacred character. In the tenth and eleventh centuries, bonzé (Buddhist priests) were sent by imperial command to recite sutras over the bodies of dead snakes found in temples. Post Wheeler, an American authority on Japanese mythology, claims that, in the remote and little-traveled interior of some of the islands, men may still be found who keep snakes in cages and offer them prayers as "honorable-lords-serpents."

Real giant snakes have existed in Japan only as imports from mainland Asia, brought to serve as patron spirits in shrines and temples, particularly in those of the fertility and love goddess, Benzai-ten. Originally the goddess was Princess Jewel-of-Store-house-Rice, a *kami* (divine or deity) daughter of He-Who-Invites and She-Who-Invites, Japan's first deity pair. Her alter ego is a white snake with the face of an aged man. Her sacred day is the day of the snake, and suppliants most easily command her attention between 9 and 11 a.m., the hours of the snake.

Benzai-ten's most important shrine is on the island of Enoshima. There, during the reign of Kimmei-Tenno (1138–1215), the goddess is said to have shown herself as a serpent 200 feet long to Hojo Tokimasa when he sought her help to insure the prosperity of his house. When she plunged into the sea, she left behind three huge scales which Hojo collected and arranged in a design for his family crest.

In Japan, white snakes, giant or otherwise, were once thought to be "Shintai," the god bodies of the Shinto deities who were themselves invisible. When the spirits of the prominent dead show themselves to the living, it is in serpent form. The imperial princess Taku-Hata, who hanged herself because she was accused of unchastity, sometimes appears as a rainbow shaped like a serpent.[56]

In 367, Tamichi, a famous warrior, appeared to a band of maurauding Ainus who had surrounded his tomb as a large venomous serpent, who bit and killed the men who intended to violate his grave.[57]

A legend from the province of Mimasaka indicates that one of its most ancient deities was a serpent, Kaya by name, to whom a virgin girl was sacrificed each year. In the eighth century, Echizen province had two great snake gods named the "Rainy-Night-Kami" and the "Great-Worm-Kami."

The Hydra killed by Hercules, Fafnir slain by the German hero Siegfried, and the multiheaded Nagas of India have their counterparts in Japanese mythology in the Eight-Forked-Serpent-of-Koshi who evidently was the guardian spirit of a voracious river with a many-mouthed delta.[58] According to the legend, a Sky-Kami named Brave-Swift-Impetuous-Male, who was born to the first creative pair, was so hyperactive and disruptive that the gods expelled him to the earth. In his wanderings he chanced upon an aged couple caressing a young girl and crying. The Kami asked to know the cause of their grief, and they told him that originally they had had eight daughters, but each year the Serpent of Koshi demanded one as a sacrifice. The child to whom they were clinging was the last.

> The Kami asked about the snake and the old man replied: Its eyes are red as the winter-cherry; its body is single, eight-headed and eight-tailed. Rock firs grow on each of its heads, on each of its sides is a mountain, and on its back grow moss, pine-trees, and cryptomeria. Its length as it crawls trails over

eight valleys and eight hills, and its belly to the sight is always bloody and fomented. It is a very fearful beast.

The Kami looked at the maiden and coveted her, and the old man said, "We will respectfully offer her to you. We pray you to first slay the serpent and afterwards it will be well to give her your favor."

Brave-Swift-Impetuous-Male distilled some "eightfold refined poisonous spirits," and poured the liquor into eight tubs. When the serpent came to claim the maiden, it straightway dipped its eight heads into the tubs and became soddenly drunk. The Kami struck off the heads and chopped the body into eight pieces. At the tail he found that something blunted his blade and, reaching in, he pulled out a fabulous sword. A copy of this blade, known as Herb-Queller, is one of the three treasures which symbolize Japan's imperial house. The others are the Sun's mirror and a necklace of ancient jewels.[59]

The Phallic Camudi

Most American Indian snake mythology is based on the rattlesnake, the real snake who becomes the fabulous feathered serpent. Of the three giant snakes in tropical America, the bushmaster, boa constrictor, and anaconda, only the last seems to have given rise to sizable body of folk literature, although Jaime Villa, a herpetologist in Nicaragua, reports that people in his country believe that boas are born of bitches and bark like dogs. Boas are also considered venomous, but only at 4 a.m. and 4 p.m. When they bathe, they are said to leave their poison in the form of a yellow gelatin on a rock near the river's edge. After bathing, they eat the poison. Villa suggests that the confusion about the boa constrictor's being venomous may result from bitten persons' mistaking poisonous bushmasters for boas.[60]

Anaconda mythology mostly comes from the Orinoco River drainage system, where the Indians claim that Purrunaminari, the supreme god, created their nations. For a while all was well, but a monstrous snake appeared to destroy and devour the people and it soon seemed that mankind would be exterminated. However, a son of the creator killed the snake, leaving its huge body near the river. As the serpent began to putrefy, large worms developed in its entrails. Each worm was transformed into a Carib Indian and his women, who, like the snake which produced them, were savage, inhuman, and cruel.

The origin of the Caribs is the theme of another myth which has at least two versions. According to the Warrau people, Warrau women were forbidden to bathe during their menses. One woman, more fastidious then the others, bathed nonetheless and attracted a large camudi (anaconda) who caught her and coupled with her. She became pregnant but concealed her condition. During the next few weeks, her brothers noticed that, although she rarely took any tools with her when she went to gather fruit and nuts, she always came back laden with choice produce. They secretly followed her into the jungle and there watched as she met her serpent lover and copulated with him. Then the snake coiled up a large nut tree, where he turned into a man and shook down the fruit for the woman. After her baskets were full, the man, transformed into his natural form, descended the tree and again made love to her. One brother said: "Something is wrong here. This will never do! My sister should not be permitted to cohabit with a snake."

The next day he took a party of his friends with him to the tree, and when the snake appeared, the men cut it into a thousand pieces. The grieving woman tenderly gathered the body of her lover together and buried it. A Carib Indian grew from each piece and the Carib nation became the sworn enemy of the Warrau people.

The Carib version differs enough in detail to make it worth relating. In it, an Indian woman and an anaconda fell in love. They met secretly every day at the bathing place. Finally, the girl's father became exasperated and called in his sons, saying, "What's the matter with your sister, why does she take so long to bathe?" The brothers followed the woman to the bathing place and saw both the camudi (anaconda) father and a baby snake being intimate with the woman. The men killed the big snake and, taking the baby far into the bush, chopped it into many pieces. Some months afterward the brothers were hunting. They came across four large huts on the spot where they had minced up the little camudi. Several people seized them and wanted to kill them, but the chief said, "These men are uncles to all of you and you must not have a bad mind toward them." The tribe was formed of Caribs, who had arisen from the pieces of the baby camudi. When the brothers returned home, they reported to their father, who immediately wished to see his grandchildren. When they returned to the encampment, the old man wept for joy to see his numerous progeny. They shared bowls of paiwarri and cassava wine and became gloriously drunk.[61]

16. Final Reflections

Today's world brings man into ever-closer contact with the larger living creatures even as their numbers dwindle almost to the vanishing point. Seldom have we felt this more keenly than during a recent trip to Grand Cayman island. Fifty feet below the rain-whipped surface of the Caribbean we drifted over an expanse of white sand between coral heads. Into this world of aqueous summer twilight swam an adult female green turtle. We took off after her, but the ingenious system of tanks, valves, and rubber fins that allowed us to enter the turtle's domain was no match for structures honed to perfection by a hundred million years of sea-turtle evolution. Even as we fell behind her, the sand below us erupted and a large eagle ray flapped off in front of us. As we sheared off to avoid it, a barracuda hung above us like a great silver lance.

A few days later as our plane circled over the southern tip of Florida, we saw some of the many islands that are the last home of the American crocodile in the United States. The water was stitched with the wakes of power boats. As we drove northward through the Florida peninsula, we transected the main part of the range of the alligator. Within a few yards of the freeway, we saw one of those saurians a good 7 feet long floating placidly in a ditch. It was probably safer there than in many other more remote places, for it would have been too risky to stop a car and shoot it.

We talked of the other big animals of Florida—the manatee,

the panther, the black bear. Even if they, along with the alligator and crocodile, could be protected from commercial exploitation and wanton killing, the time must soon come when sheer competition for space and food brings large animals into conflict with man. Florida, of course, is but a detail of a much larger picture. In Africa and Asia this competition between man and the larger wild animals is rapidly becoming acute. Hungry tribesmen and those who govern them look covetously on land set aside for wildlife. There are those who see protection of such creatures as an obstacle to personal and national aspirations. There are not many who would opt for crocodiles or pythons over people.

Implicit in human thinking is the belief that man can do something about the future of his relationship with the rest of the world fauna. At one extreme is a world in which a controlled human population will coexist with animals, most of them preserved for their own sake and surviving on man's sufferance. At the other, mankind survives at the expense of virtually every vertebrate that does not contribute directly to human welfare. But perhaps human beings overestimate both man's foresight and his power. His choices may ultimately be very limited.

A few days before this was written, we stood on the high plains of Kansas. Beneath the grass of the rolling hills lay deep beds of white shale—shale that was once the mud of a shallow Cretaceous sea. In every rain-cut gully was evidence of the life that had once swarmed in the warm, rich waters—fish vertebrae, fragments of huge clam shells, black mats of nameless plants. Here had lived plesiosaurs, mosasaurs, great gliding reptiles of the air, enormous turtles of the sea. Yet slowly—so slowly that only in the last centuries could a sentient observer have guessed what was happening—the sea disappeared and with it all the myriad creatures whose lives depended on it. Perhaps man is but one species trapped today in a drying sea, forced into ever-closer contact with his own kind and with other species whose fate is equally in the balance.

In a little pond among the shale bluffs we found the biggest reptile that lives in the region today, a large male snapping turtle. Lifted out of the water, he popped his jaws in anger. The egg from which he had hatched was buried in soil enriched by the bones of his mighty ancestors; the nucleic acids that made him a snapping turtle were not so different from those that 90 million years earlier had made *Archelon* the largest of turtles. The turtles escaped the trap of the dying Cretaceous seas; the mosasaurs and plesiosaurs did not. Turtles have escaped other traps too, and the snapper himself was designed for survival. Basically a cold-blooded, aquatic animal, he can still endure summers of searing drought and winters of savage cold. Our mammalian nucleic acids were also around in that world of the late Cretaceous, packed in the cells of non-descript, ratlike creatures, but only in the last million years has the message of the nucleotides read, "Make a man." As the snapper burrowed back in the mud of his pond, we wondered whose lineage will inherit these plains 90 million years from now.

APPENDIX

REFERENCE NOTES

GLOSSARY

SELECTED BIBLIOGRAPHY

ACKNOWLEDGMENTS

INDEX

Appendix

Geologic Time Chart of the Evolution of Reptiles

Period or Epoch	*Duration (years before present)*	*Remarks*
Recent	25,000	Reptiles reinvade formerly glaciated parts of Europe and North America.
CENOZOIC ERA		
Pleistocene	25,000 to 2 million	Reptile faunas apparently similar to those existing today. Northern faunas forced southward by glaciation.
Pliocene	2 million to 12 million	All major present-day reptile families present. Tomistomas and other crocodilians decreasing in range and variety.
Miocene	12 million to 30 million	Crocodilians widespread and diverse. Major period of snake evolution. First definite appearance of poisonous snakes.
Oligocene	30 million to 40 million	Colubrid snakes probably evolved. Oldest fossil of a venomous reptile (lizard similar to gila monster).
Eocene	40 million to 60 million	Boas and pythons the dominant snakes. Extinction of some primitive crocodilians and turtles.

Period or Epoch	*Duration (years before present)*	*Remarks*
Paleocene	60 million to 70 million	Appearance of many modern groups of lizards. Evolution and radiation of boas and other primitive snakes.
MESOZOIC ERA		
Cretaceous	70 million to 135 million	Extinction of dinosaurs and several other major reptile groups. Snakes appear near end of period. Most modern turtle families appear.
Jurassic	135 million to 180 million	Dinosaurs and other large reptiles. Primitive turtles, crocodilians, and lizards. First appearance of birds.
Triassic	180 million to 230 million	Greatest period of reptile evolution with appearance of turtles, crocodilians, dinosaurs, and numerous other groups. Mammals appear late in period.
PALEOZOIC ERA		
Permian	230 million to 275 million	First evolutionary radiation of the reptiles with extinction of most groups at end of period.
Pennsylvanian	275 million to 310 million	First reptiles evolve from amphibians near end of period.

Scientific Names of Reptiles Mentioned in the Text

CROCODILIANS AND MISCELLANEOUS GROUPS

Alligator	
American	*Alligator mississipiensis*
Chinese	*Alligator sinensis*
Caiman	
black	*Melanosuchus niger*
broad-snouted	*Caiman latirostris*
brown	*Caiman crocodilus fuscus*
dwarf	*Paleosuchus palpebrosus*
Paraguay	*Caiman crocodilus yacare*
smooth-fronted	*Paleosuchus trigonatus*
spectacled	*Caiman crocodilus crocodilus*
Crocodile	
American	*Crocodylus acutus*
African sharp-nosed	*Crocodylus cataphractus*
Cuban	*Crocodylus rhombifer*
Indian	*Crocodylus palustris*
Morlet's	*Crocodylus morleti*
New Guinea	*Crocodylus novaeguineae*
Nile	*Crocodylus niloticus*
Orinoco	*Crocodylus intermedius*
saltwater	*Crocodylus porosus*
False gharial, *see* Tomistoma	
Gharial	*Gavialis gangeticus*
Tomistoma	*Tomistoma schlegeli*
Tuatara	*Sphenodon punctatus*

TURTLES

Batagur	
Borneo	*Orlitia borneensis*

Batagur (*continued*)	
common	*Batagur baska*
painted	*Callagur pictus*
roofed	*Kachuga trivittata*
Mata-mata	*Chelys fimbriata*
Peninsula cooter	*Pseudemys floridana peninsularis*
Sea turtle	
Atlantic green	*Chelonia mydas mydas*
Atlantic ridley	*Lepidochelys kempi*
flatback	*Chelonia depressa*
hawksbill	*Eretmochelys imbricata*
leatherback	*Dermochelys coriacea*
loggerhead	*Caretta caretta*
Pacific green	*Chelonia mydas agassizi*
Pacific ridley	*Lepidochelys olivacea*
Snapping turtle	
alligator	*Macroclemys temmincki*
common	*Chelydra serpentina*
Softshell	
Bibron's	*Pelochelys bibroni*
black	*Trionyx nigricans*
Chinese	*Trionyx sinensis*
Florida	*Trionyx ferox*
Ganges	*Trionyx gangeticus*
Indian, *see* Ganges	
Indo-Chinese	*Trionyx cartilagineus*
narrow-headed	*Chitra indica*
Nile	*Trionyx triunguis*
peacock	*Trionyx hurum*
Texas	*Trionyx spiniferus emoryi*
Thirse, *see* Nile softshell	
Tortoise	
Aldabra	*Geochelone gigantea*
African spurred	*Geochelone sulcata*

Barrington Island	*Geochelone elephantopus subspecies*
Duncan Island	*Geochelone elephantopus ephippum*
Farquhar Island	*Geochelone gouffei*
Gadow's	*Geochelone gadowi*
Galapagos	*Geochelone elephantopus*
gopher	*Gopherus* species
hinge-backed	*Kinixys* species
horned	*Meiolania oweni*
leopard	*Geochelone pardalis*
Malagasy starred	*Geochelone radiata*
Mediterranean	*Testudo hermanni* and *T. graeca*
pancake	*Malacochersus tornieri*
Seychelles	*Geochelone sumeirei*
South American forest	*Geochelone denticulata*
Turtle	
Arrau	*Pondocnemis expansa*
box, *see* Eastern box turtle	
Eastern box	*Terrapene carolina*
Florida redbellied	*Pseudemys nelsoni*
Indian flap-shell	*Lissemys punctata*
map	*Graptemys geographica*
Mexican river	*Dermatemys mawi*
musk	*Sternotherus* species
spotted	*Clemmys guttata*
trunk, *see* leatherback sea turtle	

LIZARDS

American chameleon, *see* Anole	
Anole	*Anolis* species
Day gecko	*Phelsuma* species
Horned toad	*Phrynosoma* species
Goanna	
giant	*Varanus giganteus*
sand	*Varanus gouldii*

Iguana	
common	*Iguana iguana*
Galapagos land	*Conolophus subcristatus*
marine	*Amblyrhynchus cristatus*
rhinoceros	*Cyclura cornuta*
rock	*Ctenosaura* species
Kabara-goya, *see* Malayan monitor	
Komodo dragon	*Varanus komodoensis*
Lizard	
fence	*Sceloporus* species
Indian garden	*Calotes versicolor*
spinytail	*Uromastix hardwickeii*
viviparous	*Lacerta vivipara*
Monitor	
desert	*Varanus griseus*
green tree	*Varanus prasinus*
Indian	*Varanus bengalensis*
lace	*Varanus varius*
Malayan	*Varanus salvator*
Nile	*Varanus niloticus*
Pacific	*Varanus indicus*
water, *see* Malayan monitor	
Perenty, *see* giant goanna	

SNAKES

Anaconda	*Eunectes murinus*
southern	*Eunectes notaeus*
Boa	
Bahaman	*Epicrates striatus*
Cuban	*Epicrates angulifer*
emerald tree	*Corallus caninus*

Malagasy	*Acrantophis madagascariensis*
rosy	*Lichanura trivirgata*
rubber	*Charina bottae*
Boa constrictor	*Boa constrictor*
Bushmaster	*Lachesis muta*
Cobra	
Asian	*Naja naja*
Chinese	*Naja naja atra*
king	*Ophiophagus hannah*
Copperhead	*Agkistrodon contortrix*
Cottonmouth	*Agkistrodon piscivorus*
Cribo, *see* indigo snake	
Death adder	*Acanthophis antarcticus*
Dhaman	*Ptyas mucosus*
Fer-de-lance	*Bothrops atrox*
Jararacussu	*Bothrops jararacussu*
Krait, banded	*Bungarus fasciatus*
Mamba	
black	*Dendroaspis polylepis*
eastern green	*Dendroaspis angusticeps*
Mamushi	*Agkistrodon halys blomhoffi*
Python	
African	*Python sebae*
amethystine	*Liasis amethistinus*
black-headed	*Aspidites melanocephalus*
blood	*Python curtus*
Burmese	*Python molurus bivittatus*
Calabar	*Calabaria reinhardtii*
carpet	*Morelia spilotes variegata*

Python (*continued*)	
Colima, *see* Mexican dwarf python	
diamond	*Morelia spilotes spilotes*
Indian	*Python molurus molurus*
Mexican dwarf	*Loxocemus bicolor*
New Guinea green	*Chondropython viridis*
olive	*Liasis olivaceus*
Papuan	*Liasis papuanus*
reticulated	*Python reticulatus*
water	*Liasis fuscus*
Racer	
blue	*Coluber constrictor flaviventris*
Caspian	*Coluber jugularis caspius*
Rattlesnake, eastern diamondback	*Crotalus adamanteus*
Sea snake, erabu	*Laticauda semifasciata*
Snake	
Afro-Asian sand	*Psammophis schokari*
Australian brown	*Pseudonaja textilis*
bull	*Pituophis melanoleucus sayi*
cat-eyed	*Leptodeira* species
elephant-trunk	*Acrochordus javanicus*
indigo	*Drymarchon corais couperi* and other subspecies
keeled rat	*Ptyas carinatus*
milk	*Lampropeltis triangulum*
mud	*Farancia abacura*
mulga	*Pseudechis australis*
royal	*Spalerosophis atriceps*
tiger	*Notechis scutatus*
wart, *see* elephant-trunk snake	
Sumbadora, *see* Indigo snake	

Taipan	*Oxyuranus scutellatus*
Viper	
eyelash	*Bothrops schlegelii*
Gaboon	*Bitis gabonica*
jumping	*Bothrops nummifer*
Malay pit	*Agkistrodon rhodostoma*
Russell's	*Vipera russelii*
saw-scaled	*Echis* species

Reference Notes

1. Dinosaurs and Other Vanished Giants

1. Bjom Kurten, "Continental Drift and Evolution," *Scientific American*, CCXX(3) (1969), 54–64.
2. A. Gilbert Smith and A. Hallam, "The Fit of the Southern Continents," *Nature*, CCXV (1970), 139–144.
3. Edwin H. Colbert, *Dinosaurs: Their Discovery and Their World* (New York: Dutton, 1961), pp. 94–97.
4. Robert T. Bakker, "Ecology of the Brontosaurs," *Nature*, CCXXIX (1971), 172–174.
5. William L. Stokes, "Fossilized Stomach Contents of a Sauropod Dinosaur," *Science*, CXLIII (1964), 576–577.
6. L. Sprague de Camp and Catherine C. de Camp, *The Day of the Dinosaur* (Garden City, N.Y.: Doubleday, 1968), pp. 150–151.
7. Colbert, *op. cit.*, pp. 209–221.
8. de Camp and de Camp, *op. cit.*, p. 155.
9. *Ibid.*, pp. 199–209.
10. Norman D. Newell, "Crises in the History of Life," *Scientific American*, CCVIII(2) (1963), 77–92.

2. Introducing the Crocodilians

1. William D. Sill, "The Zoogeography of the Crocodilia," *Copeia*, 1 (1968), 76–88; Edwin H. Colbert, "*Sebecus*, Representative of a Peculiar Order of Fossil Crocodilia from Patagonia," *Bulletin of the American Museum of Natural History*, LXXXVII (1946), 217–270.
2. Barnum Brown, "The Largest Crocodile," *Natural History* (May 1942), 260–261.
3. Wann Langston Jr., "*Mourasuchus* Price, *Nettosuchus* Langston, and the Family Nettosuchidae (Reptilia: Crocodilia)," *Copeia*, 4 (1966), 882–885.

4. Wilfred T. Neill, *The Last of the Ruling Reptiles* (New York: Columbia University Press, 1971), pp. 118–119.
5. Raymond L. Ditmars, *Reptiles of the World* (New York: Macmillan, 1933), p. 10.
6. Neill, *op. cit.*, pp. 114–116.
7. Hugh B. Cott, "Scientific Results of an Inquiry into the Ecology and Economic Status of the Nile Crocodile (*Crocodilus niloticus*) in Uganda and Northern Rhodesia," *Transactions of the Zoological Society of London*, XXIX, Part 4 (1961), 227.
8. Clifford H. Pope, *The Reptile World* (New York: Knopf, 1955), pp. 13–15.
9. Alexander Hamilton, "Hamilton's Account of the East Indies," in John Pinkerton, *Voyages and Travels* (London: Longman, Hurst, Reese, Orme & Brown, 1811), vol. VIII, p. 454.
10. Paul E. P. Deraniyagala, *A Colored Atlas of some Vertebrates from Ceylon*, vol. 2, *Tetrapod Reptilia* (Colombo: The Ceylon Government Press, 1953), p. 34.
11. Neill, *op. cit.*, p. 428.
12. J. Stanley Gardiner, "A Crocodile on Rotuma," *Nature*, CIII (1919), 264.
13. Thomas Barbour, "The Crocodile in Florida," *Occasional Papers of the Museum of Zoology*, 131 (1923), 6 pp.
14. William T. Hornaday, "The Crocodile in Florida," *American Naturalist*, IX (1875), 498–504.
15. Don Ferdinand Colon, "The Life of Colon by His Son," in Pinkerton, *op. cit.*, vol. XII, p. 133.
16. Karl P. Schmidt, "Notes on Central American Crocodiles," *Field Museum of Natural History Publications, Zoological Series*, XXI (1924), 79–92; "Crocodile Hunting in Central America," *Chicago Natural History Museum Popular Series, Zoology*, 15 (1952), 1–23.
17. William A. Dunson, "Some Aspects of Electrolyte and Water Balance in Three Estuarine Reptiles," *Comparative Biochemistry and Physiology*, XXXII (1970), 161–174.
18. W. Gurnee Dyer, "Guelta of the Bleak Sahara," *Natural History*, LXXIV (November 1965), 36–39.
19. Livy (Titus Livius), *History of Rome*, translated by Alfred C. Schlesinger (Cambridge, Mass.: Harvard University Press, 1951), vol. 14, p. 179.

20. John Anderson, *Zoology of Egypt*, vol. I, *Reptilia and Batrachia* (London: Bernard Quaritch, 1898), p. 27.
21. Stanley S. Flower, "Notes on the Recent Reptiles and Amphibians of Egypt with a List of Species Recorded from that Kingdom," *Proceedings of the Zoological Society of London*, II (1933), 755–756.
22. Hymen Marx, "Checklist of the Reptiles and Amphibians of Egypt." Special Publication U. S. Naval Medical Research Unit Number Three (Cairo, Egypt U.A.R., 1968), 91 pp.
23. Flower, *op. cit.*, 745–746.
24. Cott, *op. cit.*, pp. 258–260.
25. Marco Polo, *The Book of Ser. Marco Polo*, translated by Henry Yule (London: John Murray, 1903), book II, pp. 76–78, annotation p. 81.
26. Henry Hamel, "Travels in Korea," in Pinkerton, *op. cit.*, vol. VII, p. 531.
27. Robert Swinhoe, "Note on Reptiles and Batrachians Collected in Various Parts of China," *Proceedings of the Zoological Society of London* (1870), 409–412.
28. Thomas Barbour, "A Note Regarding the Chinese Alligator," *Proceedings of the Academy of Natural Sciences, Philadelphia*, LXII (1910), 464–467.
29. Pope, *op. cit.*, pp. 30–31.
30. William Bartram, *Travels through North and South Carolina, Georgia, East and West Florida* (Philadelphia: James and Johnson, 1791), p. 123.
31. E. A. McIlhenny, *The Alligator's Life History* (Boston: The Christopher Publishing House, 1935), pp. 60–61.
32. Neill, *op. cit.*, p. 265.
33. Pope, *op. cit.*, pp. 13–15.
34. A. J. Barton, "Prolonged Survival of a Released Alligator in Pennsylvania," *Herpetologica*, XI (1955), 210.
35. Richard G. Pflanzer, "Physiological Response to Submergence Asphyxia of Crocodilian Circulatory Systems," *Proceedings Indiana Academy of Science*, LXXX (1971), 486.

3. Crocodilian Life Patterns

1. E. Ross Allen and Wilfred T. Neill, "The American Alligator," *Florida Wildlife Magazine* (October 1952), 8–9, 44.
2. Hugh B. Cott, "Scientific Results of an Inquiry into the Ecology and

Economic Status of the Nile Crocodile (*Crocodilus niloticus*) in Uganda and Northern Rhodesia," *Transactions of the Zoological Society of London*, XXIX, part 4 (1961), 253–255.

3. E. A. McIlhenny, *The Alligator's Life History* (Boston: The Christopher Publishing House, 1935), p. 71.
4. Wilfred T. Neill, *The Last of the Ruling Reptiles* (New York: Columbia University Press, 1971), p. 207.
5. Cott, *op. cit.*, p. 267.
6. K. S. Dharmakumarsinhji, "Mating and Parental Instinct in the Marsh Crocodile (*C. palustris Lesson*)," *Journal of the Bombay Natural History Society*, XLVII (1947), 174–176.
7. Miguel Alvarez del Toro, *Los Reptiles de Chiapas* (Tuxtla Gutierrez, Chiapas, Instituto Zoologico del Estado, 1960), p. 59.
8. C. S. Sonnini, *Travels in Upper and Lower Egypt* (London: John Stocksdale, 1790), vol. III, p. 259.
9. Charles R. LeBuff Jr., "Observations on Captive and Wild North American Crocodilians," *Herpetologica*, XIII (1957), 25–28.
10. Allen E. Greer Jr., "Crocodilian Nesting Habits and Evolution," *Fauna*, II (1971), 20–28.
11. McIlhenny, *op. cit.*, pp. 88–105.
12. Neill, *op. cit.*, pp. 23–24, 210–211.
13. Paul E. P. Deraniyagala, *A Colored Atlas of Some Vertebrates from Ceylon*, vol. 2., *Tetrapod Reptilia* (Colombo: The Ceylon Government Press, 1953), p. 33.
14. Wilfredo Fernandez, "The American Crocodile," *Muse News* (Miami Museum of Science), I (1970), 209–211.
15. Gottfried Hagmann, "Die Reptilien der Insel Mexiana," *Zoologischen Jahrbuchern*, 5 (1909), 473–504.
16. Cott, *op. cit.*, pp. 270–272.
17. William Bartram, *Travels Through North and South Carolina, Georgia, East and West Florida* (Philadelphia: James and Johnson, 1791), p. 126.
18. Cott, *op. cit.*, p. 275.
19. Deraniyagala, *op. cit.*, p. 33.
20. Cott, *op. cit.*, p. 304.
21. *Ibid.*, p. 276.
22. Wilfred T. Neill, "Notes on *Crocodylus novae-guineae*," *Copeia*, 1 (1946), 17–20.
23. Neill, *The Last of the Ruling Reptiles*, pp. 54–61, 282–283.

24. Cott, *op. cit.*, p. 282.
25. Paul L. Potous, *My Enemy, the Crocodile* (New York: Wilfred Funk, 1957), pp. 106–107.
26. Cott, *op. cit.*, pp. 283–297.
27. Michael J. Fogarty and J. David Albury, "Late Summer Foods of Young Alligators in Florida," *Proceedings 21st Annual Conference of Southeastern Game and Fish Commissioners* (1967), 220–222.
28. McIlhenny, *op. cit.*, pp. 40–57; Neill, *The Last of the Ruling Reptiles*, pp. 243–249.
29. Neill, *The Last of the Ruling Reptiles*, pp. 440–442.
30. C. R. S. Pitman, "The Length Attained by and the Habits of the Gharial," *Journal of the Bombay Natural History Society*, XXX (1924), 703.
31. Marcus A. Freiberg, *Vida de Batrachios y Reptiles Sudamericanos* (Buenos Aires: Cesarini, 1954), p. 190.
32. E. J. Fitkau, "Role of Caimans in the Nutrient Regime of Mouth-Lakes of Amazon Affluents," *Biotropica*, II (1970), 138–142.
33. Cott, *op. cit.*, pp. 302–304.
34. *Life Magazine*, June 4, 1971, pp. 70–74.
35. Neill, "Notes on *Crocodylus novae-guineae*," 17–20.
36. A. F. Abercromby, "Crocodile Burying Its Food," *Journal of the Bombay Natural History Society*, XXVIII (1922), 533.
37. Potous, *op. cit.*, pp. 123–124.
38. Walter W. Skeat, *Malay Magic* (London and New York: Macmillan, 1900), p. 292.
39. Cott, *op. cit.*, pp. 236–240.
40. Karl P. Schmidt, "Notes on South American Caimans," *Field Museum of Natural History, Zoological Series*, XII (1928), 205–231.
41. Cott, *op. cit.*, pp. 242–244.
42. Robert T. Bakker, "Ecology of the Brontosaurs," *Nature*, CCXXIX (1971), 172–174.
43. McIlhenny, *op. cit.*, pp. 30–39.
44. Clifford H. Pope, *The Reptile World* (New York: Knopf, 1955), pp. 30–31.
45. Fernandez *op. cit.*
46. Tony Pooley, "Crocodile Rearing in Zululand," *International Wildlife Magazine* (June 1970), 76–79.
47. Neill, *The Last of the Ruling Reptiles*, pp. 287–288.

48. Cott, *op. cit.*, p. 305.
49. Gaius Plinius Secundus (Pliny the Elder), *Historiae Naturalis* (The Historie of the World), translated by Philemon Holland (London, 1634), vol. I, p. 209.
50. Cott, *op. cit.*, p. 295.
51. Pliny, *op. cit.*, pp. 208–209.
52. Cott, *op. cit.*, 313–316.
53. *Ibid.*

4. Crocodilians and Man

1. P. B. M. Jackson, "Why do Nile Crocodiles Attack Boats?" *Copeia*, I (1962), 204–206.
2. Johnathan Richardson and Daniel Livingston, "An Attack by a Nile Crocodile on a Small Boat," *Copeia*, I (1962), 203–204.
3. Stanley S. Flower, "Notes on the Recent Reptiles and Amphibians of Egypt with a List of Species Recorded from That Kingdom," *Proceedings of the Zoological Society of London*, II (1933), 756.
4. George Cansdale, *Reptiles of West Africa* (London: Penguin Books, 1955), p. 82.
5. Hugh B. Cott, "Scientific Results of an Inquiry into the Ecology and Economic Status of the Nile Crocodile (*Crocodilus niloticus*) in Uganda and Northern Rhodesia," *Transactions of the Zoological Society of London*, XXIX (1961), 283–297.
6. W. D. Hubbard, "Crocodiles," *Copeia*, 165 (1927), 115–116.
7. Roger Caras, *Dangerous to Man* (Philadelphia: Chilton Books, 1964), pp. 206–207.
8. Arthur Loveridge, *Reptiles of the Pacific World* (New York: Macmillan, 1945), pp. 47–48.
9. Wilfred T. Neill, *The Last of the Ruling Reptiles* (New York: Columbia University Press, 1971), pp. 426–427.
10. Paul E. P. Deraniyagala, *A Colored Atlas of Some Vertebrates from Ceylon*, vol. 2, *Tetrapod Reptilia* (Colombo: The Ceylon Government Press, 1953), pp. 30, 32.
11. Miguel Alvarez del Toro, *Los Reptiles de Chiapas* (Tuxtla Gutierrez, Chiapas: Instituto Zoologico del Estado, 1960), pp. 57–58.
12. Paul L. Swanson, "Herpetological Notes from Panama," *Copeia*, 4 (1945), pp. 210–216.
13. William Bartram, *Travels through North and South Carolina, Georgia,*

East and West Florida (Philadelphia: James and Johnson, 1791), pp. 118–120.

14. John Edwards Holbrook, *North American Herpetology*, 2nd ed. (Philadelphia: J. Dobson, 1842), vol. II, p. 58.
15. E. A. McIlhenny, *The Alligator's Life History* (Boston: The Christopher Publishing House, 1935), p. 13.
16. Neill, *op. cit.*, pp. 259–261, has summarized four of these incidents; more recent ones are reported in the *Ocala Star Banner*, June 15, 1971, and *The Indianapolis Star*, July 22 and August 4, 1972.
17. Holbrook, *op. cit.*, p. 58.
18. E. Ross Allen gave us a transcript of the autopsy findings in this case. See also the *Orlando Sentinal*, July 3, 1957.
19. Don Antonio de Ulloa, "Ulloa's Voyage to South America," in John Pinkerton, *Voyages and Travels* (London: Longman, Hurst, Reese, Orme, & Brown, 1811), vol. XIV, pp. 410–411.
20. Walter W. Skeat, *Malay Magic* (London and New York: Macmillan, 1900), pp. 293–302.
21. James G. Frazer, *The Golden Bough* (London: Macmillan, 1922), vol. VIII, pp. 209–210.
22. Dick Bothwell, *Alligators* (St. Petersburg, Fla.: Great Outdoors Publishing Co., 1962), p. 68.
23. J. Monteath, "Catching Crocodiles," *Journal of the Bombay Natural History Society*, XXIX (1923), 300–301.
24. Cansdale, *op. cit.*, p. 84.
25. Gaius Plinius Secundus (Pliny the Elder) *Historiae Naturalis* (*The Historie of the World*), translated by Philemon Holland (London: 1634), vol. I, p. 209.
26. Cott, *op. cit.*, pp. 323–325.
27. E. Ross Allen and Wilfred T. Neill, "Increasing Abundance of the Alligator in the Eastern Portion of its Range," *Herpetologica*, V (1949), 109–112.
28. L. R. Heron, *An Economic Survey of the Crocodile Skin Industry in Papua and New Guinea* (Silver Springs, Fla.: International Crocodilian Society, 1970), pp. 8–11.
29. E. J. Fitkau, "Role of Caimans in the Nutrient Regime of Mouth-Lakes of Amazon Effluents," *Biotropica*, II (1970), pp. 138–142.
30. Eric Worrell, *Reptiles of Australia* (Sydney: Angus & Robertson, 1964), p. 4.

31. Holbrook, *op. cit.*, p. 59.
32. Marcus Freiberg, *Vida de Batrachios y Reptiles Sudamericanos* (Buenos Aires: Cesarini, 1954), p. 191.
33. "Scientific News," *American Naturalist*, XVI (1882), 533.
34. Pliny the Elder, *op. cit.*, vol. II, p. 314.
35. Alfred Brehm, *Brehms Tierleben: Lurche und Kriechtiere*, 2 vols. (Leipzig and Vienna: Bibliographisches Institut, 1912), I, 565.
36. Malcolm A. Smith, *The Fauna of British India including Ceylon and Burma. Reptilia and Amphibia*, vol. I. *Loricata, Testudines* (London: Taylor and Francis, 1931), p. 44.
37. Paul L. Potous, *My Enemy the Crocodile* (New York: Wilfred Funk, 1957), pp. 110–113.
38. Willis J. Abbot, *The Panama Canal* (London: Syndicate Publishing Co., 1914), pp. 286–288.
39. McIlhenny, *op. cit.*, p. 80.
40. Bothwell, *op. cit.*, pp. 11–12, 65–67.
41. Cott, *op. cit.*, pp. 326–327.

5. Myths about Crocodilians

1. Malcolm A. Smith, *The Fauna of British India Including Ceylon and Burma. Reptilia and Amphibia*, vol. I. *Loricata, Testudines* (London: Taylor and Francis, 1931), p. 48.
2. John Gould, "Extracts from the Letters of Dr. Henry Gould Relating to the Natural History of Western India Communicated by his Father, J. Gould, Esq., *Proceedings of the Zoological Society of London*, part XXIII (1855), pp. 39–45.
3. Hugh Wilkinson, *Sunny Lands and Seas* (London: John Murray, 1883), p. 327.
4. Walter W. Skeat, *Malay Magic* (London and New York: Macmillan, 1900), pp. 283–290.
5. Juvenal, *Satires*, translated by Jerome Mazzaro (Ann Arbor, Mich.: University of Michigan Press, 1965), p. 172.
6. John Anderson, *Zoology of Egypt*, vol. I. *Reptilia and Batrachia* (London: Bernard Quaritch, 1898), pp. 14–15, 24–27.
7. *Medicine and Pharmacy, an Informal History.* 1—*Ancient Egypt* (Bloomfield, N. J.: Schering Corp., 1955), no pagination.
8. *Ibid.*

9. Ilza Veith, "Four Thousand Years of Diabetes," *Modern Medicine*, XXXIX (November 1, 1971), 118–125.
10. Gaius Plinius Secundus (Pliny the Elder), *Historiae Naturalis* (*The Historie of the World*), translated by Philemon Holland (London: 1634), vol. II, pp. 314–315, 357, 419, 434.
11. A. S. Packard, E. S. Morse, A. Hyatt, and F. W. Putnam, "Rambles in Florida," *American Naturalist*, III (1870), 463–466.
12. Herodotus, *The Histories of Herodotus*, translated by Harry Carter (Haarlem: John Enschede en Zonen, 1958), pp. 119, 149.
13. Strabo, *Geography*, translated by W. Falconer and H. C. Hamilton (London: Henry G. Bohn, 1857), p. 256.
14. Richard Pococke, "Travels in Egypt," in John Pinkerton, *Voyages and Travels* (London: Longman, Hurst, Reese, Orme & Brown, 1811), vol. XV, p. 213.
15. James G. Frazer, *The Golden Bough* (London: Macmillan, 1922), vol. II, p. 584.
16. Luis Marden, "Madagascar, Island at the End of the Earth," *National Geographic*, CXXXII (October 1967), pp. 464–465.
17. James Sibree, *The Great African Island, Madagascar* (London: Trubner and Co., 1880), pp. 269–270, 284.
18. Stanley S. Flower, "Notes on the Recent Reptiles and Amphibians of Egypt with a List of Species Recorded from That Kingdom," *Proceedings of the Zoological Society of London*, II (1933), p. 758.
19. Frazer, *op. cit.*, vol. VIII, p. 213.
20. David Livingstone, *Missionary Travels and Researches in South Africa* (London: John Murray, 1857), p. 255.
21. W. C. Willoughby, "Notes on the Totemism of the Becwana," *Journal of the Anthropological Institute*, XXXV (1905), 300.
22. Joanodos Santos, "History of Eastern Ethiopia," in Pinkerton, *op. cit.*, vol. XVI, p. 700.
23. C. S. Sonnini, *Travels in Upper and Lower Egypt* (London: John Stocksdale, 1790), vol. III, p. 256.
24. Edward Gibbon, *History of the Decline and Fall of the Roman Empire* (Oxford, Ohio: David Christy, 1841), vol. I, pp. 309–310.
25. R. Chambers, *The Book of Days* (Edinburgh: W. & R. Chambers, 1881), p. 540.
26. Flower, *op. cit.*, p. 758.
27. Post Wheeler, *Sacred Scriptures of the Japanese* (New York: Henry Schuman Inc., 1952), p. 556.

28. W. F. A. Zimmerman, *Die Inseln des Indischen und Stillen Meeres* (Berlin: 1863), p. 55.
29. Frazer, *op. cit.*, vol. II, p. 152.
30. Wheeler, *op. cit.*, p. 556.
31. Gasparo Balbi, "Voyage to Pegu," in Pinkerton, *op. cit.*, vol. IX, p. 417.
32. Paul E. P. Deraniyagala, *A Colored Atlas of Some Vertebrates from Ceylon,* vol. 2. *Tetrapod Reptilia* (Colombo: The Ceylon Government Press, 1953), p. 34.
33. Wheeler, *op. cit.*, p. 557.
34. *Ibid.*, pp. 79–89.
35. *Ibid.*, Introduction, p. xiv.
36. T. Volker, *The Animal in Far Eastern Art* (Leiden: E. J. Brill, 1950), p. 57.
37. Frazer, *op. cit.*, vol. VIII, p. 212.
38. J. H. Holmes, "Notes on the Elema Tribes of the Papuan Gulf," *Journal of the Anthropological Institute of Great Britain and Ireland,* XXXIII (1903), 133.
39. Frazer, *op. cit.*, vol. I, p. 229.
40. *Reports Cambridge Anthropological Expedition to Torres Straits, 1901–1935* (Cambridge: Cambridge University Press), vol. V, p. 326.
41. J. H. Holmes, "Religious Ideas of the Elema Tribe of the Papuan Gulf," *Journal of the Anthropological Institute of Great Britain and Ireland*, XXXII (1902), 472.
42. Frazer, *op. cit.*, vol. X, pp. 249–250.
43. Walter E. Roth, "Animism and Folklore of the Guiana Indians," *Bureau of American Ethnology, 30th Annual Report* (1908–09), 313.
44. Don Ferdinand Colon, "The Life of Colon by His Son," in Pinkerton, *op. cit.*, vol. XII, p. 82.
45. Herman H. Ploss, Max Bartels, and Paul Bartels, *Woman* (London: William Heineman, 1935), p. 652.
46. John Mackay, "Maori Forest Lore," *Proceedings of the New Zealand Institute*, XL (1907), 195.
47. E. S. Craighill Handy, "Polynesian Religion," *Bernice P. Bishop Museum Bulletin 34* (1927), 128.
48. W. D. Westervelt, *Hawaiian Legends of Volcanoes* (Boston: Ellis Press, 1916), pp. 122–125.
49. William Wyatt Gill, *Myths and Songs from the South Pacific* (London: Henry S. King & Co., 1876), pp. 290–293.

50. Volker, *op. cit.*, pp. 55–66.
51. Engelbert Kaempfer, "History of Japan," in Pinkerton, *op. cit.*, vol. VII, p. 699.
52. Volker, *op. cit.*, p. 56.
53. Kaempfer, *op. cit.*, p. 699.
54. Arthur Urbane Dilley, *Oriental Rugs and Carpets*, revised by Maurice S. Dimand (Philadelphia and New York: J. B. Lippincott, 1959), p. 208.

6. Turtles around the World

1. J. F. Willgohs, "Occurrence of the Leathery Turtle in the Northern North Sea and off Western Norway," *Nature*, CLXXIX (1957), 163–164.
2. J. Sherman Bleakney, "Reports of Marine Turtles from New England and Eastern Canada," *Canadian Field-Naturalist*, LXXIX (1965), 120–128.
3. I. B. MacAskie and C. R. Forrester, "Pacific Leatherback Turtles (*Dermochelys*) off the Coast of British Columbia," *Copeia*, 3 (1962), 646.
4. N. Mrosovsky and Peter C. H. Pritchard, "Body Temperatures of *Dermochelys coriacea* and Other Sea Turtles," *Copeia*, 4 (1971), 624–631.
5. Bleakney, *op. cit.*
6. Terrence H. Leary, "A Schooling of Leatherback Turtles, *Dermochelys coriacea coriacea*, on the Texan Coast," *Copeia*, 3 (1957), 232.
7. Peter R. Bacon, "Studies of the Leatherback Turtle, *Dermochelys coriacea* (L.) in Trinidad, West Indies," *Biologia Conservatoria*, II (1970), 213–217.
8. Louis Agassiz, *Contribution to the Natural History of the United States* (Boston: Little Brown, 1857), vol. I, p. 324.
9. Archie F. Carr, *Handbook of Turtles* (Ithaca, N. Y.: Comstock Publishing Associates, 1952), p. 446.
10. Paul E. P. Deraniyagala, *A Colored Atlas of Some Vertebrates from Ceylon* (Colombo: The Ceylon Government Press, 1953), pp. 10–12.
11. Bacon, *op. cit.*

12. Archie F. Carr, "Sea Turtles: a Vanishing Asset," *International Union for the Conservation of Nature*, XIII (1968), 162–168.
13. Carlton Ray and C. W. Coates, "Record and Measurements of a Leatherback Turtle from the Gulf of Maine," *Copeia*, 3 (1958), 220–221.
14. Peter C. H. Pritchard, "Galapagos Sea Turtles: Preliminary Findings," *Journal of Herpetology*, V (1971), 1–9.
15. *Ibid.*
16. Carr, *Handbook of Turtles*, p. 385.
17. Archie F. Carr, *So Excellent a Fishe* (Garden City, N.Y.: Natural History Press, 1967), p. 226.
18. Carr, *Handbook of Turtles*, pp. 377–378.
19. Archie F. Carr and Larry Ogren, "The Ecology and Migrations of Sea Turtles, 4—The Green Turtle in the Caribbean Sea," *Bulletin of the American Museum of Natural History*, CXXI (1960), 10–21; Carr, *So Excellent a Fishe*, pp. 25–72; H. Robert Bustard, "Turtle Biology at Heron Island," *Australian Natural History*, XV (1966), 262–264.
20. Carr, *So Excellent a Fishe*, p. 186.
21. Archie F. Carr, "The Case for Long-range Chemoreceptive Piloting in *Chelonia*." In Sidney R. Galler *et al.*, eds., *Animal Orientation and Navigation* (Washington, D. C., National Aeronautics and Space Administration, 1972), pp. 469–483.
22. Howard A. Baldwin, "Long-range Tracking of Sea Turtles and Polar Bear—Instrumentation and Preliminary Results." In *ibid.*, pp. 19–28.
23. Gaius Plinius Secundus (Pliny the Elder), *Historiae Naturalis* (*The Historie of the World*), translated by Philemon Holland (London: 1634), vol. I, p. 241.
24. Woodes Rogers, *A Cruising Voyage Round the World* (London: Andrew Bell, 1718), p. 276.
25. Clarence H. Shockley, "Herpetological Notes for Ras Jiunri, Baluchistan," *Herpetologica*, V (1949), 121.
26. Archie F. Carr, "The Passing of the Fleet," *American Institute for Biological Sciences Bulletin*, October 1954, pp. 17–19.
27. Carr, *So Excellent a Fishe*, pp. 114–157.
28. Humberto Chavez, "Tagging and Recapture of the Lora Turtle," *International Turtle and Tortoise Society Journal*, III (1969), 14–19, 32–37.

29. *Ibid.*
30. Peter C. H. Pritchard, "To Find the Ridley . . . ," *International Turtle and Tortoise Society Journal*, I (1967), 30–33.
31. James A. Oliver, "An Aggregation of Pacific Sea Turtles," *Copeia*, 2 (1946), 103.
32. Pliny, *op. cit.*, vol. I, p. 242.
33. Carr, *Handbook of Turtles*, pp. 371–373.
34. Bruce Halstead, *Poisonous and Venomous Marine Animals of the World* (Washington, D. C.: U. S. Government Printing Office, 1965–1970), vol. III, pp. 617–627.
35. Oliver, *op. cit.*
36. Carr, *Handbook of Turtles*, pp. 388–389.
37. *The Indianapolis Star*, May 21, 1972.

7. The Big Freshwater Turtles

1. Nelly de Rooij, *The Reptiles of the Indo-Australian Archipelago*, vol. I, *Lacertilia, Chelonia, Emydosauria* (Leiden: E. J. Brill, 1915), p. 331.
2. Clifford H. Pope, *The Reptiles of China, Natural History of Central Asia*, vol. X (New York: The American Museum of Natural History, 1935), p. 57.
3. Malcolm A. Smith, *The Fauna of British India including Ceylon and Burma, Reptilia and Amphibia*, vol. I *Loricata, Testudines* (London: Taylor and Francis, 1931), p. 157.
4. B. L. Chaudhuri, "Aquatic Tortoises of the Middle Ganges and Brahmaputra," *Records of the Indian Museum*, VII (1912), 212–214.
5. Malcolm A. Smith, *op. cit.*, p. 164.
6. Stanley S. Flower, "Notes on the Recent Reptiles and Amphibians of Egypt with a List of Species Recorded from That Kingdom," *Proceedings of the Zoological Society of London*, II (1933), p. 755.
7. Nelson Annandale and M. H. Shastri, "Relics of the Worship of Mud-Turtles (Trionychidae) in India and Burma," *Journal and Proceedings of the Asiatic Society of Bengal*, X (1914), 131–138.
8. Guy Mountfort, *The Vanishing Jungle* (Boston: Houghton Mifflin, 1970), pp. 93–94.

9. Robert G. Webb, "North American Recent Soft-shelled Turtles (Family Trionychidae)," *University of Kansas Publications of the Museum of Natural History*, XIII (1962), 563.
10. Flower, *op. cit.*, p. 754.
11. Chaudhuri, *op. cit.*
12. Pope, *op. cit.*, p. 58.
13. Robert Snedigar, *Our Small Native Animals, Their Habits and Care* (New York: Dover Publications Inc., 1963), p. 206.
14. James L. Dobie, "Reproduction and Growth in the Alligator Snapping Turtle, *Macroclemys temmincki, Copeia*, 4 (1971), 645–658.
15. Philip W. Smith, "The Amphibians and Reptiles of Illinois, *Illinois Natural History Survey Bulletin*, 28 (1961), 121–122.
16. M. M. Wickham, "Notes on the Migration of *Macrochelys lacertina*," *Proceedings of the Oklahoma Academy of Science*, 2 (1922), 20–22.
17. Roger Conant, director, Philadelphia Zoo, personal communication.
18. E. Ross Allen and Wilfred T. Neill, "The Alligator Snapping Turtle in Florida," *Ross Allen Reptile Institute, Special Publication No. 4*, pp. 1–15.
19. Donald A. Hammer, "The Durable Snapping Turtle," *Natural History*, LXXX (1971), 59–64.
20. Archie F. Carr, *Handbook of Turtles* (Ithaca, N. Y.: Comstock Publishing Associates, 1952), p. 63.
21. Hammer, *op. cit.*
22. Karl P. Schmidt and Robert F. Inger, *Living Reptiles of the World* (Garden City, N. Y.: Hanover House, 1957), p. 15.
23. Alvin R. Cahn, "The Turtles of Illinois," *Illinois Biological Monographs*, XVI (1937), 41.
24. *The Indianapolis Star*, March 9–20, March 25, April 12, 1949.
25. Malcolm A. Smith, *op. cit.*, p. 153.
26. Peter C. H. Pritchard, *Living Turtles of the World* (Jersey City, N. J.: T. F. H. Publications, 1967), p. 25.
27. de Rooij, *op. cit.*, pp. 291–293.
28. John Mehrtens, "*Orlitia* the Bornean Terrapin," *International Turtle and Tortoise Society Journal*, IV (1970), 6–7, 33.
29. Pritchard, *op. cit.*, pp. 221–222.
30. Henry Walter Bates, *A Naturalist on the River Amazon* (2 vols. London: John Murray, 1863), vol II, pp. 212–213, 270–273.

31. Marcos A. Freiberg, *Vida de Batracios y Reptiles Sudamericanos* (Buenos Aires: Cesarini, 1954), pp. 178–179.
32. J. J. C. Mallinson, "The River Turtles of the Amazon," *International Turtle and Tortoise Society Journal*, I (1966), 34–35.
33. Robert C. Lee, "Observing the Tortuga Blanca," *International Turtle and Tortoise Society Journal*, III (1969), 32–34.

8. Giant Land Tortoises

1. Ray Pawley, "The Hidden Tortoise of Torreon," *International Turtle and Tortoise Society Journal*, II (1968), 20–23, 36.
2. David Porter, *A Voyage in the South Seas* (London: Sir Richard Phillips and Co., 1823), p. 46.
3. Peter C. H. Pritchard, *Living Turtles of the World* (Jersey City, N. J.: T. F. H. Publications, 1967), p. 17.
4. *Ibid.*, pp. 156–188.
5. John Van Denburgh, "The Gigantic Land Tortoises of the Galapagos Archipelago," *Proceedings of the California Academy of Sciences, Fourth Series*, II (1914), 215–228.
6. *Ibid.*, p. 228.
7. Peter C. H. Pritchard, "A Further Report on Galapagos Tortoises," *Herpetological Review*, III (1971), 25–27, 49–91.
8. Pritchard, *Living Turtles of the World*, pp. 173–177.
9. *Ibid.*, pp. 148–153.
10. Charles Darwin, *Journal of Researches into the Natural History and Geology of the Countries Visited During the Voyage of H. M. S. Beagle* (New York: The Heritage Press, 1957), p. 341.
11. *Ibid.*, pp. 349–350.
12. Ramon P. Noegel, "Scanning the Indian Ocean Islands," *International Turtle and Tortoise Society Journal*, I (1967), 30–36, 44–45.
13. Karl P. Schmidt and Robert F. Inger, *Living Reptiles of the World* (Garden City, N. Y.: Hanover House, 1957), p. 29.
14. Pritchard, *Living Turtles of the World*, p. 181.
15. Schmidt and Inger, *op. cit.*, p. 29.
16. Pritchard, *Living Turtles of the World*, p. 181.
17. Terry Graham and Victor Hutchison, "Centenarian Box Turtles," *International Turtle and Tortoise Society Journal*, III (1969), 25–29.

18. Van Denburgh, *op. cit.*, pp. 333–334.
19. Darwin, *op. cit.*, p. 348.
20. Noegel, *op. cit.*, pp. 35, 36.
21. William Dampier, *A New Voyage Round the World* (London, James Knapton, 1697), p. 102.
22. Noegel, *op. cit.*, pp. 31, 32.
23. Van Denburgh, *op. cit.*, p. 243.
24. Pritchard, "A Further Report on Galapagos Tortoises," *op. cit.*, 25–27, 49–51.

9. Legendary Turtles

1. Gaius Plinius Secundus (Pliny the Elder), *Historiae Naturalis* (*The Historie of the World*), translated by Philemon Holland (London, 1634), vol. I, pp. 431, 433.
2. *Ibid.*, pp. 431–432.
3. *Ibid.*
4. *Ibid.*
5. Henry G. Fischer, "Egyptian Turtles," *Bulletin of the Metropolitan Museum of Art*, XXIV (February 1966), 195.
6. Quoted in *ibid.*, p. 196.
7. Geraldine Marshall and Frank du Pre, "Turtles in Art and Legend, Part V: Mythology, Folklore and Superstition," *International Turtle and Tortoise Society Journal*, III (1) (1969), 16.
8. Arthur Basham, "Hinduism," *Encyclopaedia Britannica* (Chicago: William Benton, 1964), vol. XI, p. 507.
9. T. Volker, *The Animal in Far Eastern Art* (Leiden: E. J. Brill, 1950), p. 171.
10. Nelson Annandale and M. H. Shastri, "Relics of the Worship of Mud-Turtles (Trionychidae) in India and Burma," *Journal and Proceedings of the Asiatic Society of Bengal*, X (1914), 131–138.
11. *Ibid.*
12. Volker, *op. cit.*, p. 172.
13. Leonid Skvirsky, "Turtles in Art and Legend, Part IV: Lord of the Northern Quadrangle," *International Turtle and Tortoise Society Journal*, II(3) (1968), 22.
14. Volker, *op. cit.*, pp. 171–173.

15. *Ibid.*, p. 172.
16. Katherine M. Ball, *Decorative Motives in Oriental Art* (London: John Lane: The Bodley Head, 1927), p. 49.
17. Volker, *op. cit.*, p. 171.
18. Ball, *op. cit.*, p. 51.
19. Post Wheeler, *The Sacred Scriptures of the Japanese* (New York: Henry Schuman, 1952), pp. 290–293.
20. Quoted in *ibid.*, p. 542.
21. Englebert Kaempfer, "History of Japan," in John Pinkerton, *Voyages and Travels* (London: Longman, Hurst, Reese, Orme & Brown, 1811), vol. VII, pp. 709–710.
22. Clifford H. Pope, *The Reptile World* (New York: Knopf, 1955), p. 75.
23. Ball, *op. cit.*, pp. 41–42.
24. *Walam Olum: The Migration Legend of the Lenni Lenape or Delaware Indians* (from the original translation by Constantine Rafinesque) (Indianapolis: Indiana Historical Society, 1954), pp. 48–49.
25. Cosmos Mindeleff, "Navaho Houses," *17th Annual Report of the Bureau of American Ethnology 1895–96* (1898), part 2, pp. 488–489.
26. *Ibid.*
27. Ellen R. Emerson, *Indian Myths* (Boston, 1884), p. 119.
28. *Ibid.*
29. James G. Frazer, *Totemism and Exogamy* (London: Macmillan, 1910), vol. I, p. 579.
30. *Ibid.*, vol. III, p. 23.
31. J. N. B. Hewitt, "Iroquoian Cosmology," *21st Annual Report of the Bureau of American Ethnology 1899–1900* (1903), p. 360.
32. Cottie Burland, *North American Indian Mythology* (Feltham, Middlesex, England: The Hamlyn Publishing Group Ltd., 1968), pp. 61–62.

10. The Great Lizards

1. Bjom Kurten, "Continental Drift and Evolution," *Scientific American*, CCXX (3) (1969), 54–64.
2. Alfred S. Romer, *Osteology of the Reptiles* (Chicago: University of Chicago Press, 1956), p. 559.

3. Thomas Barbour, "Defense Posture of *Varanus gouldii*," *Copeia*, 1 (1943), 56–57.
4. David Stammer, "Goannas," *Wildlife in Australia*, VII (1970), 118–120.
5. James A. Kern, "Dragon Lizards of Komodo," *National Geographic*, CXXXIV (6) (December 1968), 875.
6. Walter Auffenberg, "A Day with Number 19: Report on a Study of the Komodo Monitor," *Animal Kingdom*, LXXIII (1970), 19–23.
7. Douglas Burden, "The Quest for the Dragon of Komodo," *Natural History*, XXVII (1) (1927), 3–18.
8. Paul E. P. Deraniyagala, *A Colored Atlas of Some Vertebrates from Ceylon*, vol. 2, *Tetrapod Reptilia* (Colombo: The Ceylon Government Press, 1953), pp. 86–87.
9. Auffenberg, *op. cit.*, p. 19.
10. *Ibid.*, p. 21; Karl P. Schmidt and Robert F. Inger, *Living Reptiles of the World* (Garden City, N.Y.: Hanover House, 1957), p. 170.
11. Auffenberg, *op. cit.*, pp. 20–23.
12. Deraniyagala, *op. cit.*, p. 87.
13. Auffenberg, *op. cit.*, p. 22.
14. George Cansdale, *Reptiles of West Africa* (London: Penguin Books, 1955), p. 63.
15. Deraniyagala, *op. cit.*, p. 84.
16. Raymond L. Ditmars, *Reptiles of the World* (New York: Macmillan, 1933), p. 95.
17. Clifford H. Pope, *The Reptile World* (New York, Knopf, 1955), p. 311.
18. Zdenek Vogel, *Reptiles and Amphibians: Their Care and Behaviour*, English edition translated and revised by Gwynne Vevers (London: Studio Vista, 1964), p. 140.
19. Walter Auffenberg, "Komodo Dragons," *Natural History*, LXXXI (April 1972), 54–57.
20. Harold Cogger, *Australian Reptiles in Colour* (Sydney: A. H. and A. W. Reed, 1967), p. 52.
21. Deraniyagala, *op. cit.*, p. 86.
22. Schmidt and Inger, *op. cit.*, p. 170.
23. Alfred Brehm, *Brehms Tierleben Lurche und Kriechtiere*, vol. II (Leipzig and Vienna: Bibliographisches Institut, 1913), p. 130.
24. Walter W. Skeat, *Malay Magic* (New York: Macmillan, 1900), p. 289.
25. C. S. Sonnini, *Travels in Upper and Lower Egypt* (London: John Stocksdale, 1790), vol. III, p. 200.

26. David F. Avery and Wilmer W. Tanner, "Evolution of Iguanine Lizards (Sauria, Iguanidae) as Determined by Osteological and Myological Characters," *Brigham Young University Science Bulletin*, XII (1971), 1–79.
27. Charles C. Carpenter, "The Marine Iguana of the Galapagos Islands, its Behavior and Ecology," *Proceedings of the California Academy of Sciences*, XXXIV (1966), 329–376.
28. Edmund S. Hobson, "Observations on Diving in the Galapagos Marine Iguana," *Copeia*, 2 (1965), 249–250.
29. Charles Darwin, *Journal of Researches into the Natural History and Geology of the Countries Visited During the Voyage of H. M. S. Beagle* (New York: The Heritage Press, 1957), pp. 351–353.
30. *Ibid.*, p. 353.
31. Pope, *op. cit.*, p. 239.
32. Paul Swanson, "The Iguana *Iguana iguana iguana* (L)" *Herpetologica*, VI (1950), 191.
33. A. Stanley Rand, "A Nesting Aggregation of Iguanas," *Copeia*, 3 (1968), 552–561.
34. Charles C. Carpenter, "Behavioral Studies on Reptiles—Bobs, Nods, and Pushups," *The American Biology Teacher*, XXVIII (1966), 527–529.
35. Swanson, *op. cit.*, p. 190.
36. *Ibid.*, p. 192.
37. David Porter, *A Voyage in the South Seas* (London: Sir Richard Phillips & Co., 1823), p. 40.
38. Swanson, *op. cit.*, pp. 187–193.

11. Snakes: The Big Five and Their Relatives

1. James A. Oliver, *Snakes in Fact and Fiction* (New York: Macmillan, 1959), p. 10.
2. Ramona Morris and Desmond Morris, *Men and Snakes* (New York: McGraw-Hill, 1965), p. 120.
3. *Awake!*, March 8, 1971.
4. C. R. S. Pitman, *A Guide to the Snakes of Uganda* (Kampala, Uganda: The Uganda Society, 1938), pp. 219–220.
5. Richard Estes, T. H. Frazzetta, and Ernest Williams, "Studies on the Fossil Snake *Dinilysia patagonica* Woodward: Part I. Cranial

Morphology," *Bulletin of the Museum of Comparative Zoology*, CXL (1970), 25–73.

6. Clifford H. Pope, *The Giant Snakes* (New York: Knopf, 1965), pp. 164–165.
7. Alfred S. Romer, *Osteology of the Reptiles* (Chicago: University of Chicago Press, 1956), pp. 562–563, 571.
8. Charles M. Bogert, "Boas—a Paradoxical Family," *Animal Kingdom*, LXXII (1969), 18–25.
9. Oliver, *op. cit.*, pp. 22–26.
10. Roger Caras, *Dangerous to Man* (Philadelphia: Chilton Books, 1964), p. 136.
11. James A. Peters and Braulie Orejas-Miranda, "Catalogue of the Neotropical Squamata: Part I, Snakes," *U. S. National Museum Bulletin 297* (1970), 114–115.
12. Oliver, *op. cit.*, pp. 30–31.
13. Stanley S. Flower, "Notes on a Second Collection of Reptiles Made in the Malay Peninsula and Siam from November 1896, to September 1898," *Proceedings of the Zoological Society of London for 1899*, p. 655.
14. *Ibid.*, p. 654.
15. Malcolm A. Smith, *The Fauna of British India including Ceylon and Burma. Reptilia and Amphibia*, vol. III, *Serpentes* (London: Taylor and Francis, 1943), p. 110.
16. Simon Campden-Main, *A Field Guide to the Snakes of South Vietnam* (Washington, D. C.: Smithsonian Institution Press, 1970), p. 10.
17. *The News, Mexico City*, September 14, 1971.
18. *The Congressional Record*, April 6, 1971, E2867.
19. Pope, *op. cit.*, p. 21.
20. Jean Doucet, "Les Serpents de la République de Côte d'Ivoire," *Acta Tropica*, XX (1963), 227.
21. Pope, *op. cit.*, p. 158.
22. Gaius Plinius Secundus (Pliny the Elder), *Historiae Naturalis* (*The Historie of the World*), translated by Philemon Holland (London, 1634), vol. I, book 8, p. 199.
23. Frank Wall, *Snakes of Ceylon* (Colombo: H. R. Cottle, Government Printer, 1929), p. 69.
24. Strabo, *Geography*, translated by Horace L. Jones (London: William Heineman, 1917), vol. VII, p. 79.

25. *Dawn* (Karachi), February 20, 1960.
26. Pope, *op. cit.*, pp. 30–31.
27. Eric Worrell, *Reptiles of Australia* (Sydney: Angus and Robertson, 1964), p. 97.
28. T.H. White, *The Book of Beasts, being a Translation from a Latin Bestiary of the Twelfth Century* (London: Jonathan Cape, 1954), p. 180.
29. Peters and Orejas-Miranda, *op. cit.*, pp. 37–38.
30. Oliver, *op. cit.*, p. 36.
31. Laurence M. Hardy and Roy W. McDiarmid, "The Amphibians and Reptiles of Sinaloa, Mexico," *University of Kansas Publications of the Museum of Natural History*, XVIII (1969), 155–156.
32. Karl P. Schmidt and Robert F. Inger, *Living Reptiles of the World* (Garden City, N. Y.: Hanover, 1957), p. 177.
33. Worrell, *op. cit.*, pp. 98–100.
34. Nelly de Rooij, *The Reptiles of the Indo-Australian Archipelago*, Vol. II *Ophidia* (Leiden: E. J. Brill, 1917), pp. 19–25.

12. Biology and Behavior of Boas and Pythons

1. Walter Rose, *Snakes—Mainly South African* (Capetown: Maskew Miller Ltd., 1955), pp. 82–83.
2. James Sibree, *The Great African Island, Madagascar* (London: Trubner, 1880), p. 56.
3. William Bosman, "Bosman's Guinea, The Gold, Slave, and Ivory Coasts," in John Pinkerton, *Voyages and Travels* (London: Longman, Hurst, Reese, Orme & Brown, 1811), vol. XVI, p. 448.
4. Clifford H. Pope, *The Giant Snakes* (New York: Knopf, 1965), p. 155.
5. G. Kingsley Noble, "The Sense Organs Involved in the Courtship of *Storeria, Thamnophis* and Other Snakes," *Bulletin of the American Museum of Natural History*, LXXIII (1937), 673–725.
6. Pope, *op. cit.*, pp. 119–121.
7. *Ibid.*, p. 122.
8. *Ibid.*, pp. 128–147.
9. Allen Vinegar, Victor Hutchison, and Herndon Dowling, "Metabolism, Energetics, and Thermoregulation during Breeding of Snakes of the Genus *Python*," *Zoologica*, LV (1970), 19–47.
10. Wilfred T. Neill, "Viviparity in Snakes: Some Ecological and Zoogeo-

graphical Considerations," *American Naturalist*, XCVIII (1964), 35–55.

11. Henry S. Fitch, "Reproductive Cycles in Lizards and Snakes," *University of Kansas Museum of Natural History Miscellaneous Publications 52* (1970), 116–118.
12. Robert W. Hanlon, "Reproductive Activity of the Bahaman Boa (*Epicrates striatus*), *Herpetologica*, XX (1964), 143–144.
13. Wilfred T. Neill, "Parturient Anaconda, *Eunectes gigas* Latreille, Eating Its Own Abortive Eggs and Foetal Membranes," *Quarterly Journal of the Florida Academy of Science*, XXV (1962), 73–75.
14. Frank Wall, *Snakes of Ceylon* (Colombo: H. R. Cottle, Government Printer, 1921), p. 60.
15. Laurence M. Hardy and Roy W. McDiarmid, "The Amphibians and Reptiles of Sinaloa, Mexico," *University of Kansas Publications of the Museum of Natural History*, XVIII (1969), 155–156.
16. Felix Kopstein, "Ein Beitrag zur Eierkunde und sur Fortpflanzung der malaiischen Reptilien," *Bulletin of the Raffles Museum*, XIV (1938), 165.
17. Vivian Fitzsimons, *Snakes of Southern Africa* (Capetown: Purnell and Sons, 1962), pp. 96–97.
18. Pope, *op. cit.*, p. 123; Fitch, *op. cit.*, p. 118.
19. William Beebe, *Jungle Days* (New York: G. P. Putnam's Sons, 1925), pp. 3–25.
20. William Beebe, "Field Notes on the Snakes of Kartabo, British Guiana and Caripito, Venezuela," *Zoologica*, XXXI, part I (1946), 11–52.
21. Eric Worrell, *Reptiles of Australia* (Sydney: Angus and Robertson, 1964), p. 96.
22. Ludolf Wehekind, "Notes on the Food of Trinidad Snakes," *British Journal of Herpetology*, II (1955), 9–13.
23. Worrell, *op. cit.*, p. 99.
24. James A. Oliver, *Snakes in Fact and Fiction* (New York: Macmillan, 1959), p. 43.
25. Pope, *op. cit.*, p. 86.
26. Rose, *op. cit.*, p. 77.
27. Oliver, *op. cit.*, p. 43.
28. Albert Schwartz and Larry H. Ogren, "A Collection of Reptiles and Amphibians from Cuba with Descriptions of Two New Forms," *Herpetologica*, XII (1956), 91–110.

29. Robert Barrett, "The Pit Organs of Snakes," in Carl Gans and Thomas Parsons, eds., *The Biology of the Reptilia* (New York: Academic Press, 1971), vol. II, pp. 289–298.
30. Oliver, *op. cit.*, pp. 74–75.
31. Rose, *op. cit.*, p. 77.
32. Pope, *op. cit.*, p. 160.
33. *Ibid.*, p. 156.
34. Herndon G. Dowling, "Boa Constrictor: from Tropical Menace to Popular Pet," *Animal Kingdom*, LXVIII (1965), 183–185.
35. Pope, *op. cit.*, pp. 85, 162–163.
36. Rose, *op. cit.*, pp. 80–81.
37. Pope, *op. cit.*, p. 227.
38. Wall, *op. cit.*, p. 60.
39. Rose, *op. cit.*, pp. 79–80.
40. C. J. P. Ionides, *Mambas and Man-Eaters, a Hunter's Story* (New York: Holt, Rinehart and Winston, 1965), p. 228.
41. R. M. Isemonger, *Snakes of Africa, Southern, Central and East* (Johannesburg: Thomas Nelson and Sons, 1962), p. 165.
42. Arthur Loveridge, "On Two Amphibious Snakes of the Central African Lake Region," *Bulletin of the Antivenin Institute of America*, V (1931), 7–12.
43. Joan Root and Alan Root, "Mzima, Kenya's Spring of Life," *National Geographic*, CXL (3) (1971), 350–373.
44. Henry Walter Bates, *A Naturalist on the River Amazons* (London: John Murray, 1863), vol. II, p. 114.
45. Oliver, *op. cit.*, p. 48.
46. Pope, *op. cit.*, p. 228.
47. Oliver, *op. cit.*, pp. 48–49.
48. Charles Mayer, *Trapping Wild Animals in Malay Jungles* (New York: Duffield, 1921), p. 87.
49. James Clarke, *Man Is the Prey* (New York: Stein and Day, 1969), pp. 193–194.
50. *Tampa Tribune*, March 16, 1961.
51. C. R. S. Pitman, *A Guide to the Snakes of Uganda* (Kampala, Uganda: The Uganda Society, 1938), p. 85.
52. Laurence M. Klauber, *Rattlesnakes* (Berkeley, California: University of California Press, 1956), vol. II, pp. 1033–1079.
53. Isemonger, *op. cit.*, p. 17.

54. Pitman, *op. cit.*, p. 220.
55. Pope, *op. cit.*, p. 176.
56. *Ibid.*, p. 175.
57. Klauber, *op. cit.*, pp. 1033–1036.
58. Isemonger, *op. cit.*, pp. 12–13.
59. Pope, *op. cit.*, p. 235.
60. Lee R. Wolin, J. M. Ordy, and Arline Dillman, "Monkeys' Fear of Snakes: a Study of its Basis and Generality," *Journal of Genetic Psychology*, CIII (1963), 207–226.
61. Jules H. Masserman and Curtis Pechtel, "Neuroses in Monkeys: A Preliminary Report of Experimental Observations," *Annals of the New York Academy of Sciences*, LVI (1953), 253–265.

13. Other Large Snakes

1. C. A. Domergue, J. Richaud, and E. R. Brygoo, "Application des techniques serologiques a l'etude de la systematique des serpents de Madagascar" (Application of serological techniques to the study of the classification of Madagascar snakes), *Comptes Rendus des Séances de Société Biologie*, CLXIV (1970), 2690–2692.
2. Garth Underwood, *A Contribution to the Classification of Snakes* (London: The British Museum (Natural History), 1967) particularly pp. 47–57, 71–80, and 142–145.
3. James R. Dixon, "A Large Bullsnake, *Pituophis catenifer sayi*, from Texas," *Copeia*, 3 (1952), 193.
4. Laurence M. Klauber, *Rattlesnakes* (Berkeley, Calif.: University of California Press, 1956), vol. I, pp. 145–147.
5. William E. Duellman, "Amphibians and Reptiles of the Rainforests of Southern El Peten, Guatemala," *University of Kansas Publications of the Museum of Natural History*, XV (1963), 239.
6. Karl P. Schmidt and Robert F. Inger, *Living Reptiles of the World* (Garden City, N. Y.: Hanover House, 1957), p. 210.
7. Frank Wall, *Snakes of Ceylon* (Colombo: H. R. Cottle, Government Printer, 1921), p. 184.
8. Edward H. Taylor, "The Serpents of Thailand and Adjacent Waters," *University of Kansas Science Bulletin*, XLV(9) (1965), 957.
9. C. J. Aagaard, "Cobras and King Cobras," *Natural History Society of Siam Journal*, VI (1924), 315–316.

10. Raymond L. Ditmars, *Snakes of the World* (New York: Macmillan, 1931), p. 150.
11. M. W. F. Tweedie, *The Snakes of Malaya* (Singapore: Government Printing Office, 1954), p. 100.
12. H. Alistair Reid, "Symptomatology, Pathology, and Treatment of Land Snake Bite in India and Southeast Asia," in Wolfgang Bücherl, Eleanor Buckley, and Venancio Deulofeu, eds., *Venomous Animals and Their Venoms* (New York: Academic Press, 1968), vol. I, p. 629.
13. Somsri Ganthavorn, "A Case of King Cobra Bite," *Toxicon*, IX (1971), 293–294.
14. Harry Kursh, *Cobras in His Garden* (Irvington-on-Hudson, N. Y.: Harvey House, 1965), pp. 126–133.
15. James A. Oliver, *Snakes in Fact and Fiction* (New York: Macmillan, 1959), p. 126.
16. Jonothan H. E. Leakey, "Observations Made on King Cobras in Thailand during May, 1966," *Journal of the National Research Council of Thailand*, V (1969), 1–10.
17. Walter Rose, *Snakes—Mainly South African* (Capetown: Maskew Miller, 1955), p. 104.
18. C. J. P. Ionides, *Mambas and Man-Eaters, A Hunter's Story* (New York: Holt, Rinehart, and Winston, 1965), p. 152.
19. Andre Villiers, "La Collection de Serpents de l'IFAN (Acquisitions 1953)," *Bulletin de l'Institut Français d'Afrique Noire*, XVI (1954), 1244–1245.
20. Ionides, *op. cit.*, pp. 163–165.
21. David S. Chapman, "The Symptomatology, Pathology, and Treatment of the Bites of Venomous Snakes of Central and Southern Africa," in Bücherl *et al., op. cit.,* vol. I, pp. 477–481.
22. Ionides, *op. cit.*, pp. 155–157.
23. Schmidt and Inger, *op. cit.*, p. 238.
24. Vivian Fitzsimons, *Snakes of Southern Africa* (Capetown: Purnell and Sons, 1962), pp. 310–311.
25. *Ibid.*, p. 311.
26. George Cansdale, *Reptiles of West Africa* (London: Penguin Books, 1955), pp. 48–49; E. Grasset, "La Vipre du Gabon," *Acta Tropica*, III (1946), 97–115.
27. Eric Worrell, *Reptiles of Australia* (Sydney: Angus and Robertson, 1964), p. 133.

28. H. Flecker, "More Fatal Cases of Bites of the Taipan (*Oxyuranus scutellatus*)," *Medical Journal of Australia*, II (1944), 383–384; C. H. Campbell, "'The Taipan (*Oxyuranus scutellatus*) and the Effect of Its Bite," *Medical Journal of Australia*, I (1967), 735–769.
29. Campbell, *op. cit.*
30. Worrell, *op. cit.*, p. 133.
31. Bayard Brattstrom, "The Fossil Pit-Vipers (Reptilia Crotalidae) of North America," *Transactions of the San Diego Society of Natural History*, XII (1954), 36–37.
32. Carolus Linnaeus, *Systema Naturae*. 12th rev. ed. (Stockholm, 1766), vol. I, pp. 372–373.
33. Alfred Brehm, *Brehms Tierleben: Lurche und Kriechtiere* (Leipzig and Vienna: Bibliographisches Institut, 1913), vol. II, p. 545.
34. Raymond L. Ditmars, *Reptiles of the World* (New York: Macmillan, 1933), p. 244.
35. Afranio do Amaral, "Lista remissiva dos Ophidios do Brasil," *Memorias do Instituto Butantan*, IV (1929), 112.
36. Quoted in Karl P. Schmidt, "Anent the 'Dangerous' Bushmaster," *Copeia*, 3 (1957), 233.
37. Oliver, *op. cit.*, pp. 186–187.
38. Brehm, *op. cit.*, p. 547.
39. Ditmars, *Reptiles of the World*, p. 245.
40. Sherman A. Minton Jr. and Madge R. Minton, *Venomous Reptiles* (New York: Charles Scribner's Sons, 1969), pp. 75–77.
41. Donald A. Jutzy, Stanley H. Biber, Norman W. Elton, and Earl C. Lowry, "A Clinical and Pathological Analysis of Snake Bites on the Panama Canal Zone," *American Journal of Tropical Medicine and Hygiene*, II (1953), 129–141.

14. Snakes as Leather, Food, and Medicine

1. Clifford H. Pope, *The Giant Snakes* (New York: Knopf, 1965), pp. 240–241.
2. *The Snake,* vols. I–III, 1969–1971, advertisements; also Richard C. Goris, "Extraordinary Feeding Behavior of *Rhabdophis tigrinus* under Semi-captive Conditions," in *ibid.*, II (1970), 133–134.
3. Robert E. Kuntz, *Snakes of Taiwan* (Taipei, Taiwan: U. S. Medical Research Unit No. 2, 1963), p. 77.

4. Pope, *op. cit.*, p. 242; also *The Indianapolis Star*, October 9, 1965, and November 21, 1971.
5. Ronald M. Berndt and Catherine H. Berndt, *The World of the First Australians* (Sydney: Ure Smith, 1964), p. 228.
6. Ramona Morris and Desmond Morris, *Men and Snakes* (New York: McGraw-Hill, 1965), p. 18.
7. Marcus A. Freiberg, *Vida de Batrachios y Reptiles Sudamericanos* (Buenos Aires: Cesarini, 1954), p. 91.
8. Kuntz, *op. cit.*, pp. 77–78.
9. Katherine M. Ball, *Decorative Motives of Oriental Art* (London: The Bodley Head, 1927), p. 173.
10. Walter W. Skeat, *Malay Magic* (London and New York: Macmillan, 1900), pp. 302–303.
11. Paul Weinstein, Henry Krawczyk, and James Peers, "Sparganosis in Korea," *American Journal of Tropical Medicine and Hygiene*, III (1954), 112–127.
12. K. Prathap, K. S. Lau, and J. M. Bolton, "Pentastomiasis: A Common Finding at Autopsy Among Malaysian Aborigines," *American Journal of Tropical Medicine and Hygiene*, XVIII (1969), 20–27.

15. Snake Mythology

1. *The Indianapolis Star*, February 8, 1972.
2. James G. Frazer, *The Worship of Nature* (New York: Macmillan, 1926), p. 596–598.
3. Wilfrid D. Hambly, "The Serpent in African Belief and Custom," *Journal of the American Anthropological Association*, new series, 31 (1929), 663.
4. Frazer, *op. cit.*, pp. 596–598.
5. Jesse W. Fewkes, "Tusayan Katchinas," *15th Annual Report of the Bureau of American Ethnology 1893–'94* (1897), pp. 271–272.
6. Emerson F. Greenman, *Guide to Serpent Mound* (Columbus, Ohio: The Ohio Historical Society, 1964), pp. 4, 9–13, 17–18.
7. Verrier Elwin and Geoffery Cumberlege, *The Muria and Their Ghotul* (Oxford: Oxford University Press, 1947), p. 262.
8. Katherine M. Ball, *Decorative Motives in Oriental Art* (London: John Lane; The Bodley Head Ltd., 1927), pp. 175–176.
9. George Turner, *Samoa* (London: Macmillan, 1884), p. 300.

10. James G. Frazer, *Belief in Immortality* (London, Macmillan, 1924), vol. III, p. 263.
11. Matilda Coxe Stevenson, "The Sia," *11th Annual Report of the Bureau of American Ethnology* (1894), 69, 124–126.
12. *Walam Olum: The Migration Legend of the Lenni Lenape or Delaware Indians* (from the original translation by Constantine Rafinesque) (Indianapolis: Indiana Historical Society, 1954), pp. 35–49.
13. *Ibid.*, pp. 209–215.
14. *Arizona Daily Star*, Tucson, Arizona, December 15, 1968.
15. H. A. Guerber, *Myths of Northern Lands* (New York: American Book Co., 1895), pp. 19–21, 264–270.
16. P. Grappin, "Germanic Lands: The Mortal Gods," in Pierre Grimal, ed., *Larousse World Mythology* (New York: Prometheus Press, 1965), pp. 378–380.
17. Guerber, *op. cit.*, pp. 264–270.
18. H. A. Rey, *The Stars* (Boston: Houghton Mifflin, 1952), pp. 48, 52.
19. Joseph Fontenrose, *Python* (Berkeley, Calif.: University of California Press, 1959), p. 15.
20. *Ibid.*, pp. 13–14.
21. Ernest A. Gardner and Eugene Vanderpool, "Delphi," *Encyclopaedia Britannica* (Chicago: William Benton, 1964), vol. VII, p. 208.
22. Fontenrose, *op. cit.*, p. 72.
23. Pierre Grimal, "Greece, Myth and Logic," in Pierre Grimal, ed., *Larousse World Mythology*, p. 123.
24. Abbé Barthelemy, *Travels of Anacharsis the Younger in Greece* (Philadelphia: Jacob Johnson, 1804), pp. 12–17.
25. M. Oldfield Howey, *The Encircled Serpent* (London: Rider & Co., 1928), pp. 146–147.
26. James G. Frazer, *Totemism and Exogamy* (London: Macmillan, 1910), vol. II, pp. 585–586.
27. *Ibid.*, p. 585.
28. *Ibid.*, pp. 585–586.
29. William Bosman, "Bosman's Guinea, The Gold, Slave, and Ivory Coasts," in John Pinkerton, *Voyages and Travels* (London: Longman, Hurst, Reese, Orme, & Brown, 1811), vol. XVI, pp. 494–500.
30. James G. Frazer, *The Golden Bough* (London: Macmillan, 1925), vol. VIII, p. 288.
31. *Ibid.*, pp. 289–290.

32. James Sibree, *The Great African Island, Madagascar* (London: Trubner, 1880), p. 276.
33. *Ibid.*
34. Frazer, *The Golden Bough*, vol. V, p. 41.
35. *Ibid.*, vol. X, p. 31.
36. C. R. S. Pitman, *A Guide to the Snakes of Uganda* (Kampala, Uganda: The Uganda Society, 1938), p. 215.
37. F. W. Fitzsimons, *The Snakes of South Africa* (Capetown: T. Maskew Miller, 1919), pp. 65–69.
38. Roland Robinson, *The Feathered Serpent* (Sydney: Edwards and Shaw, 1956), p. 56.
39. Sherman A. Minton Jr. and Madge Rutherford Minton, *Venomous Reptiles* (New York: Charles Scribner's Sons, 1969), p. 192.
40. Ronald M. Berndt and Catherine H. Berndt, *The World of the First Australians* (Sydney: Ure Smith, 1964), p. 154.
41. *Ibid.*
42. *Ibid.*, pp. 148, 211.
43. Robinson, *op. cit.*, pp. 30–35.
44. C. F. Gordon-Cumming, *At Home in Fiji* (Edinburgh & London: William Blackwood & Sons, 1882), p. 348.
45. *Ibid.*, pp. 349–350.
46. Anthony Alpers, *Legends of the South Seas* (New York: Thomas Y. Crowell, 1970), p. 371.
47. *Ibid.*, pp. 73–75.
48. Minton and Minton, *op. cit.*, pp. 131, 132, 136.
49. T. Watley [untitled communication], *Asiatick Researches or Transactions of the Society Instituted in Bengal*, III (1792), 343.
50. Frank Outram, "The King Cobra Performs," *Travel*, LXIII(1) (1934), pp. 14–17, 55, 56.
51. Harry Miller, "The Cobra, India's 'Good Snake' " *National Geographic*, CXXXVIII (3) (1970), pp. 393–409.
52. Mircea Eliade, *Cosmos and History* (New York: Harper and Row, 1959), p. 19.
53. Elwin and Cumberlege, *op. cit.*, p. 253.
54. J. A. MacCulloch, "Serpent Worship," in James Hastings, ed., *Encyclopedia of Religion and Ethics* (New York: Charles Scribner's Sons, 1908, 1927), vol. XI, p. 407.
55. Fitzsimons, *op. cit.*, pp. 46, 48.

56. Post Wheeler, *Sacred Scriptures of the Japanese* (New York: Henry Schuman, 1952), pp. 402–406.
57. T. Volker, *The Animal in Far Eastern Art* (Leiden: E. J. Brill, 1950), p. 144.
58. Wheeler, *op. cit.*, p. 406.
59. *Ibid.*, pp. 33–37.
60. Jaime Villa, *Serpientes Venenosas de Nicaragua* (Managua, Nicaragua, [n.p.] 1962), pp. 78–80.
61. Walter E. Roth, "Animism and Folk-Lore of the Guiana Indians," *30th Annual Report of the Bureau of American Ethnology 1908–1909* (1915), 143–144.

Glossary

ANTIVENIN (ANTIVENENE, ANTIVENOM): An antiserum produced against a venom and capable of neutralizing its effects.

BASILISK: A mythical reptile with a deadly glance, said to be hatched from the egg of a cock.

BIOMASS: The amount of living matter present in a particular habitat.

BIOTA: Plants and animals of a particular geographic region.

CARAPACE: The upper shell of a turtle.

CARIB: American Indian tribes living in northern Brazil, the Guianas, Venezuela, Colombia, and the Caribbean.

CHELONIAN: Pertaining to turtles or tortoises; a member of either group.

CHEMOSENSOR: An organ or structure for detecting chemical stimuli such as taste or smell.

CLOACA: The common opening of the intestinal, urinary, and female reproductive tracts in reptiles.

COCKATRICE: A mythical reptile said to have the legs and wings of a cock and the head of a serpent.

DYAK: Indonesian people in the interior region of Borneo.

ELAPID: Member of a family of venomous snakes characterized by having fixed or slightly movable fangs in the anterior part of the upper jaw.

FLAGELLATES: Single-celled organisms with flagella or whiplike locomotor structures.

HERPETOLOGIST: A zoologist specializing in the study of reptiles and amphibians.

JOGI: Snake charmers of India and Pakistan.

LEVIATHAN: Biblical sea monster which symbolizes evil, probably based on the crocodile.

LITTORAL: Shore zone between high and low watermarks.

MAUND: A Hindi unit of weight, about 80 pounds.

NAGA: Snake gods of ancient India.

NEOTROPICAL: Pertaining to the American or New World tropics.

OCCIPITAL SHIELDS: A pair of large scales at the base of the head in some snakes and lizards.

OPHIOLATRY: The worship of snakes.

PELAGIC: Pertaining to or living in the open sea.

PLASTRON: The lower shell of a turtle.

POLYVALENT: Containing antibodies specific for several substances or organisms. Applied to ANTIVENINS produced against venoms of several species of snakes.

PREMAXILLARY TEETH: Teeth on the premaxillary bone which is in the extreme front of the jaw in snakes.

RAPTORIAL: Adapted to seize prey: usually applied to birds such as hawks and owls.

SAURIAN: Pertaining to lizards and crocodilians; a reptile belonging to either of these groups.

SEXUAL DIMORPHISM: Differences between the sexes involving features other than the sexual organs.

SOLFUGID: A harmless spiderlike desert arachnid, also known as wind scorpion or sun scorpion.

Selected Bibliography

ANDERSON, JOHN. *Zoology of Egypt.* Vol. I: *Reptilia and Batrachia.* London: Bernard Quaritch, 1898.

AUFFENBERG, WALTER. "Komodo Dragons," *Natural History,* LXXXI (April 1972), pp. 52–59.

BATES, HENRY WALTER. *A Naturalist on the River Amazon.* 2 vols. London: John Murray, 1863.

BELLAIRS, ANGUS. *The Life of Reptiles.* 2 vols. New York: Universe Books, 1970.

BERNDT, RONALD M., and BERNDT, CATHERINE H. *The World of the First Australians.* 1st Australian ed. Sidney: Ure Smith, 1964.

BOTHWELL, DICK. *Alligators.* St. Petersburg, Fla.: Great Outdoors Publishing Co., 1962.

BRONGERSMA, L. D. *European Atlantic Turtles.* Leiden: E. J. Brill, 1972.

CANSDALE, GEORGE. *Reptiles of West Africa.* London: Penguin Books, 1955.

CARAS, ROGER. *Dangerous to Man.* Philadelphia: Chilton Books, 1964.

CARR, ARCHIE F. *Handbook of Turtles.* Ithaca, N.Y.: Comstock Publishing Associates, 1952.

———. *So Excellent a Fishe.* Garden City, N. Y.: Natural History Press, 1967.

CARR, ARCHIE F., and OGREN, LARRY. "The Ecology and Migrations of Sea Turtles," *Bulletin of the American Museum of Natural History,* CXXI (1960).

COLBERT, EDWIN H. *Dinosaurs, Their Discovery and Their World.* New York: E. P. Dutton, 1961.

COOPER, W. R. *The Serpent Myths of Ancient Egypt.* London: Robert Hardwicke, 1873.

COTT, HUGH B. "Scientific Results of an Inquiry into the Ecology and Economic Status of the Nile Crocodile (*Crocodilus niloticus*) in Uganda and northern Rhodesia," *Transactions of the Zoological Society of London,* XXIX, part 4 (1961), 211–337.

DERANIYAGALA, P. E. P. *A Colored Atlas of Some Vertebrates from Ceylon,* vol. II, *Tetrapod Reptilia.* Colombo: The Ceylon Government Press, 1953.

DE CAMP, L. SPRAGUE, and DE CAMP, CATHERINE C. *The Day of the Dinosaur.* Garden City, N. Y.: Doubleday, 1968.

DE ROOIJ, NELLY. *The Reptiles of the Indo-Australian Archipelago.* Vols. I and II. Leiden: E. J. Brill, 1915–1917.

DITMARS, RAYMOND L. *Reptiles of the World.* New York: Macmillan, 1933.

EMERSON, ELLEN RUSSELL. *Indian Myths.* Minneapolis: Ross & Haines, 1965.

FITZSIMONS, VIVIAN. *Snakes of Southern Africa.* Cape Town: Purnall & Sons, 1962.

FLOWER, STANLEY S. "Notes on a Second Collection of Reptiles Made in the Malay Peninsula and Siam from November, 1896 to September, 1898," *Proceedings of the Zoological Society of London for 1899*, pp. 600–697.

———. "Notes on the Recent Reptiles and Amphibians of Egypt with a List of Species Recorded from that Kingdom," *Proceedings of the Zoological Society of London*, 1933, vol. 2, pp. 735–851.

FRAZER, JAMES G. *The Golden Bough.* London: Macmillan, 1922.

FREIBERG, MARCUS A. *Vida de Batrachios y Reptiles Sudamericanos.* Buenos Aires: Cesarini, 1954.

HOWEY, M. OLDFIELD. *The Encircled Serpent.* London: Rider, 1928.

IONIDES, C. J. P. *Mambas and Man-Eaters, A Hunter's Story.* New York: Holt, Rinehart and Winston, 1965.

ISEMONGER, R. M. *Snakes of Africa, Southern Central and East.* Johannesburg: Thomas Nelson and Sons, 1962.

KERN, JAMES A. "Dragon Lizards of Komodo," *National Geographic*, CXXXIV (1968), pp. 872–880.

KEARY, CHAS. F. *Outlines of Primitive Belief Among the Indo-European Races.* London: Longmans, Green, 1882.

KING, F. WAYNE. "Adventures in the Skin Trade," *Natural History*, LXXX (1971), p. 8.

KURTEN, BJOM. "Continental Drift and Evolution," *Scientific American*, CCXX (1969), pp. 54–64.

MCILHENNY, E. A. *The Alligator's Life History.* Boston: The Christopher Publishing House, 1935.

MERTENS, ROBERT. *The World of Amphibians and Reptiles.* English translation by H. W. Parker. London: George G. Harrap, 1960.

MORRIS, RAMONA, and MORRIS, DESMOND. *Men and Snakes.* New York: McGraw-Hill, 1965.

NEILL, WILFRED T. *The Last of the Ruling Reptiles.* New York: Columbia University Press, 1971.

NEWELL, NORMAN D. "Crises in the History of Life," *Scientific American,* 208 (1963), pp. 77–92.

OLIVER, JAMES A. *Snakes in Fact and Fiction.* New York: Macmillan, 1959.

POPE, CLIFFORD H. *The Reptile World.* New York: Alfred A. Knopf, 1955.

———. *The Giant Snakes.* New York: Alfred A. Knopf, 1965.

PRITCHARD, PETER C. H. *Living Turtles of the World.* Jersey City, N. J.: T. F. H. Publications, 1967.

RAYMOND, PERCY E. *Prehistoric Life.* Cambridge, Mass.: Harvard University Press, 1939.

RIVETT-CARNAS, J. H. "The Snake Symbol in India," *Journal of the Asiatic Society,* Bengal, 1881.

ROMER, ALFRED S. *Osteology of the Reptiles.* Chicago: University of Chicago Press, 1956.

ROSE, WALTER. *Snakes—Mainly South African.* Capetown: Maskew Miller, 1955.

SCHMIDT, KARL P., and INGER, ROBERT F. *Living Reptiles of the World.* Garden City, N. Y.: Hanover House, 1957.

SIBREE, JAMES. *The Great African Island, Madagascar.* London: Trubner, 1880.

SKEAT, WALTER W. *Malay Magic.* New York: Macmillan, 1900.

SWANSON, PAUL. "The Iguana *Iguana iguana iguana*," *Herpetologica,* VI (1950), pp. 187–193.

UNDERWOOD, GARTH. *A Contribution to the Classification of Snakes.* London: British Museum (Natural History), 1967.

VAN DENBURGH, JOHN. *The Gigantic Land Tortoises of the Galapagos Archipelago.* Proceedings of the California Academy of Sciences, Fourth Series, II, 1914.

VOGEL, JEAN PHILLIPE. *Indian Serpent Lore.* London: Arthur Probsthain, 1926.

WALL, FRANK. *Snakes of Ceylon.* Colombo: H. R. Cottle, Government Printer, 1921.

WORRELL, ERIC. *Reptiles of Australia.* Sydney: Angus & Robertson, 1964.

Acknowledgments

Our thanks are due to many people who helped us in the preparation of this book. Friends and colleagues in herpetology include Dr. Roger Conant, Director, Philadelphia Zoo; Itzchak Gilboa and Dr. Herndon Dowling, Herpetological Information Search Systems, American Museum of Natural History; Ross Allen, Andrew Koukoulis, William Gleason, and David Ibbetson, Ross Allen Reptile Institute; Dr. Phillip W. Smith, Illinois Natural History Survey; Dr. James A. Peters and Robert Tuck, Department of Herpetology, U. S. National Museum; Dr. Ilya S. Darevsky, U. S. S. R. Academy of Sciences, Leningrad Museum; Eddie M. Ashmore, Kentucky Herpetological Society; Craig E. Pinkus and Larry Morris, Indianapolis, Indiana. We are also specially indebted to Mike Bowman and our friends in the British Sub-Aqua Club, Indianapolis; Dr. H. B. Bechtel, Valdosta, Georgia; Dr. J. Alan Holman, Michigan State University; Joan Hamilton Morris, Rare Books Librarian, and Richard A. Davis, Librarian, Irwin Library, Butler University; Maryellen Brezie, Josiah Q. Bennett, and Constance Work, Lilly Library, Indiana University; Rebecca Dixon, Head Librarian, Institute for Sex Research, Indiana University; Mary Jane Laatz and her staff, Indiana University Medical Center Library; Marie Crockett, Shews Wilson, and April Minton. For their help with translations we thank Brooks and Miguel Cervantes, Queretaro, Mexico, and Helen Pilgrim-Minor, Indianapolis.

Index

About the Authors

Sherman A. Minton, Jr., was born in Indiana and received his M.D. from Indiana University in 1942. He is professor of microbiology at the School of Medicine of Indiana University. A noted herpetologist who has contributed to numerous scientific and technical journals, Dr. Minton has served as president of the International Society of Toxinology. He is a Fellow and Council Member of the Herpetologists League and Research Associate in Herpetology of the American Museum of Natural History.

Madge Rutherford Minton, also born in Indiana, studied at Medill School of Journalism and Indiana University and received her A.B. from Butler University in 1941. She has assisted her husband with various procedures, particularly the extracting of snake venom, and has worked closely with him in the field.